Animal Populations

A Study of Physical, Conceptual, and Mathematical Models

Second Edition

A Problem–Based Unit Designed
for 6th–8th Grade Learners

CENTER FOR GIFTED EDUCATION · THE COLLEGE OF WILLIAM & MARY

KENDALL/HUNT PUBLISHING COMPANY
4050 Westmark Drive   Dubuque, Iowa 52002

Book Team
Chairman and Chief Executive Officer: *Mark C. Falb*
President and Chief Operating Officer: *Chad M. Chandlee*
Director of National Book Program: *Paul B. Carty*
Editorial Development Manager: *Georgia Botsford*
Developmental Editor: *Lynnette M. Rogers*
Vice President, Operations: *Timothy J. Beitzel*
Assistant Vice President, Production Services: *Christine E. O'Brien*
Senior Production Editor: *Charmayne McMurray*
Permissions Editor: *Renae Horstman*
Cover Designer: *Jenifer Chapman*

Author Information for Correspondence and Workshops:
Center for Gifted Education
The College of William and Mary
P.O. Box 8795
Williamsburg, VA 23187-8795
Phone: 757-221-2362
Email address: *cfge@wm.edu*
Web address: *www.cfge.wm.edu*

Center for Gifted Education Staff, First Edition
Project Director: Dr. Joyce VanTassel-Baska
Project Managers: Dr. Shelagh A. Gallagher
Dr. Victoria B. Damiani
Project Consultants: Dr. Beverly T. Sher
Linda Neal Boyce
Dana T. Johnson
Dr. Donna L. Poland
Unit Developers: Dana T. Johnson
Dr. Beverly T. Sher

Center for Gifted Education Staff, Second Edition
Executive Director: Dr. Joyce VanTassel-Baska
Director: Dr. Elissa F. Brown
Curriculum Director and Unit Editor: Dr. Kimberley L. Chandler
Curriculum Writer: Emily Johnson
Curriculum Reviewers: Susan McGowan
Glen Stevens

The William and Mary Center for Gifted Education logo is a depiction of the Crim Dell Bridge, a popular site on the William and Mary campus. Since 1964 this Asian-inspired structure has been a place for quiet reflection as well as social connections. The bridge represents the goals of the Center for Gifted Education: to link theory and practice, to connect gifted students to effective learning experiences, to offer career pathways for graduate students, and to bridge the span between general education and the education of gifted learners.

Contents

Introduction

Unit Introduction

Animal Populations is a problem-based science unit designed for high-ability learners that has been used successfully with students in a wide variety of settings, from pull-out programs for high-ability learners to traditional heterogeneously grouped classrooms. It allows middle school students to learn population biology in a novel way, namely through the process of grappling with an ill-structured, "real-world" problem.

Because the unit is problem-based, the way a teacher implements the unit will differ from the ways most traditional science units are used. Preparing for and implementing problem-based learning requires time, flexibility, and a willingness to experiment with teaching strategies. This unit is complex in terms of science, mathematics, and technology content. Teachers who do not have the required expertise should identify experts who can assist them. Since middle schools often encourage teaming, teachers are encouraged to take advantage of this practice in the implementation of the unit.

Rationale

Animal Populations has been designed to introduce middle grade level students to population biology in an engaging fashion. The problem-based learning format was chosen in order to allow students to acquire significant science content knowledge in the course of solving an interdisciplinary, real world problem. This format requires students to analyze the problem situation, to discover and select information that applies to the problem solution, and to obtain that information in a variety of ways. In addition to research, students, with teacher facilitation, will conduct experiments of their own design in order to generate useful data and to formulate possible problem solutions. This problem-based method allows students to model the scientific process, from the problem-finding and information-gathering steps to the evaluation of experimental data and the recasting or solution of the problem. Finally, the overarching scientific concept of models provides students with a framework for the analysis of both their experiments and the problem as a whole.

Implementation Time

The total time required for completion of *Animal Populations* is a minimum of 40 hours, with more time required for additional activities.

Assessment

Animal Populations contains many assessment opportunities that can be used to monitor student progress and assess student learning. Opportunities for formative assessment:

- The student's Problem Log is a written compilation of the student's thoughts about the problem. Most lessons contain suggested questions for students to answer in their Problem Logs. The Problem Log should also be used by the student to record and store data and new information that have been obtained during the course of the unit.

- Other metacognitive forms are used to help the student explain his/her solutions to particular parts of the problem.

- Teacher observation of student participation in large- and small-group activities is another opportunity for ongoing assessment.

Opportunities for summative assessment include:

- The final resolution activity involves a small-group presentation of a solution for the unit's ill-structured problem. The quality of the solution will reflect the group's understanding of the science involved as well as the societal and ethical considerations needed to form an acceptable solution.

- Final post-assessments allow the teacher to determine whether individual students have met the systems objectives, science content, and science process skills listed in the Curriculum Framework at the beginning of the unit.

Appendix E, Suggested Rubrics, includes suggestions for assessing experimental design process skills, connections to the systems concept, oral presentations, and persuasive writing.

New to This Edition

Student books are available for purchase to provide students an opportunity to record information about the problems as they progress through the units.

Included in the student books are the Problem Log Questions, Student Brainstorming Guides, Experimental Design Diagrams, Student Experiment Protocols, and Student Laboratory Reports, along with the Glossary and Laboratory Safety Precautions.

The books are designed to be consumable.

Words to the Wise Teacher:

The unit you are about to begin, *Animal Populations,* consists of 19 lessons; the unit requires a minimum of 40 hours of instruction. The end of this section of the unit contains a letter for parents that you may wish to send home with your students or use as a template for your own letter to be distributed before beginning the unit. The letter describes the goals of the curriculum as well as ways parents can supplement the unit at home.

The unit includes many opportunities for students to participate actively in solving a real-world problem. Some of these activities involve homework that supplements class work; others involve research conducted in a library/media center or online. Please read the unit before beginning to teach so that you have a sense of when you might need materials and assistance from your media specialist.

Handouts for the unit are included, as well as some background information on various topics. A separate notebook or Problem Log is required for each student. A materials list at the beginning of each lesson notes specific items for that lesson; however, you may wish to procure additional items depending upon the outcomes of sessions with your students.

Several methods for assessing student progress are indicated in the unit. Assessments ask students to demonstrate understanding of the unit concept within the relevant context. Writing activities will include essays, a research project, and Problem Log responses throughout the unit. Post-assessments are included that may be used to compare student achievement at the end of the unit to their knowledge at the beginning.

A section detailing some implementation guidelines and the key teaching models of the unit follows the lesson plans. Teachers are encouraged to read this section and, if possible, to attend an implementation workshop on the units. Contact the Center for Gifted Education for more information.

The Center for Gifted Education thanks you for your interest in our materials!

Alignment to National Council of Teachers of Mathematics Standards

	Animal Populations	NCTM Standards
Math Content	1. Students will be able to estimate the size of an animal population using various survey techniques. 2. Students will understand the principle of exponential growth. 3. Students will be able to understand and use a mathematical model that accurately represents the size and growth of an animal population and factors that affect its change over time. 4. Students will be able to graph and analyze population data.	Students will: 1. Select methods for solving problems, including estimation. 2. Understand properties of functions including exponential functions. 3. Model and solve problems using various representations such as graphs and equations.

Alignment to National Science Education Standards and Benchmarks for Science Literacy

Category of Standard	*Animal Populations*	National Science Education Standards	Benchmarks for Science Literacy
Concept	Students will be able to: 1. Use and create physical, conceptual, and mathematical models.	Models are tentative schemes or structures that correspond to real objects, events, or classes of events, and that have explanatory power. Models help scientists and engineers understand how things work. Models take many forms, including physical objects, plans, mental constructs, mathematical equations, and computer simulations.	Models are often used to think about processes that happen too slowly, too quickly, or on too small a scale to observe directly, or that are too vast to be changed deliberately, or that are potentially dangerous. Choosing a useful model is one of the instances in which intuition and creativity come into play in science, mathematics, and engineering.
Science Content Topics	Science: Students will be able to: 1. Use population biology terminology. 2. Understand the biological niche that deer occupy and the ecosystem of which they are a part. Technology: Students will be able to: 1. Become familiar with the use of a graphing calculator (such as the TI-83) to graph data and to investigate exponential growth. 2. Become familiar with technological research tools such as the Internet.	Life Science: All students should develop understanding of populations and ecosystems. Science in Personal and Social Perspective: All students should develop understanding of populations, resources, and environments.	Life Science: In all environments organisms with similar needs may compete with one another for resources, including food, space, water, air, and shelter. In any particular environment, the growth and survival of organisms depend on the physical conditions. Technology: Students should: 1. Use calculators to compare amounts proportionally. 2. Use computers to store and retrieve information.

Alignment to National Science Education Standards and Benchmarks for Science Literacy *continued*

Category of Standard	*Animal Populations*	National Science Education Standards	Benchmarks for Science Literacy
Scientific Processes	Students will be able to: 1. Identify meaningful scientific problems for investigation. 2. Use appropriate scientific techniques to address the scientific problems they have identified. These include data gathering, observation, experimentation, mathematical modeling, communication, and data analysis.	Students should be able to: 1. Identify questions that can be answered through scientific investigations. 2. Design and conduct a scientific investigation. 3. Use appropriate tools and techniques to gather, analyze, and interpret data. 4. Develop descriptions, explanations, predictions, and models using evidence.	Students should: 1. Organize information in simple tables and graphs and identify relationships they reveal. 2. Locate information in reference books, back issues of newspapers and magazines, compact disks, and computer databases. 3. Find and describe locations on maps using coordinates. 4. Write clear, step-by-step instructions for conducting investigations, operating something, or following a procedure.

Models Outcomes

Goal 1: To understand the concept of models
Students will be able to use and analyze several models during the course of the unit. They will create, test, and apply physical, conceptual, and mathematical models.

A. Students will use and create physical, conceptual, and mathematical models that will be used to aid in the solution of the problem posed by the unit.
   1. Students will use physical models to understand the nature of the deer problem in the community. They will also create their own physical models, such as maps showing the relative levels of deer damage and use these models in the process of creating a solution to the problem posed in the unit.
   2. Students will use and create conceptual models to understand the methods used by population biologists. They will plan experiments using these conceptual models, which will help them to test various approaches to solving the problem posed by the unit.
   3. Students will use and create mathematical models to predict the changing size of the deer population over time as well as to test the effectiveness of various forms of human intervention in controlling the deer population.
B. Students will justify the creation of each model made in the course of the unit by explaining the potential utility of the model in finding a solution or in understanding the problem.
C. Students will test their model's ability to predict the attributes of the object or process modeled by comparing the predictions of the model with real data about what was modeled.
D. Students will refine their models to enhance their ability to predict the behavior of modeled processes or objects.

Content Goal and Outcomes

Goal 2: To use the interdisciplinary principles of population biology
Students will become familiar with scientific, mathematical, and technological methods and information relevant to the areas of population biology, ecology, and human health.

Specific Content Outcomes

Science
A. Students will become familiar with a new scientific area, namely population biology.
B. Students will understand the biological niche that deer occupy and the ecosystem of which they are a part.
C. Students will be able to identify and evaluate deer population control measures.
D. Students will be able to describe the cause, effects, symptoms, treatments, and preventative measures for Lyme disease.

Mathematics
A. Students will be able to estimate the size of an animal population using various survey techniques.
B. Students will understand the principle of exponential growth.
C. Students will be able to understand and use a mathematical model that accurately represents the size and growth of an animal population and factors that affect its change over time.
D. Students will be able to graph and analyze population data.

Technology
A. Students will become familiar with the use of a graphing calculator to graph data and to investigate exponential growth.
B. Students will become familiar with technological research tools such as the Internet.

continued

Process Goals and Outcomes

Goal 3: To use scientific process skills to identify and solve problems.

In order to solve these scientific problems, students will be able to design, perform, and report on the results of a number of experiments, observations, and calculations.

Scientific Process Outcomes

A. Students will be able to identify meaningful scientific problems for investigation during the course of working through the population problem and its ramifications. These problems may include determining the local incidence of Lyme disease, finding out how many deer are present in the problem locale and determining the extent of their impact on the local community, assessing the nature of the response of local residents to deer impact, investigating the effectiveness of various deer damage control measures, such as deer repellents.

B. Students will be able to use appropriate scientific techniques to address the scientific problems they have identified. These include observation (including observation of habits of deer, surveys), experimentation (testing the effectiveness of deer repellents) and calculation (mathematical modeling and data analysis).

C. During their scientific studies, students will be able to:
   - Demonstrate appropriate data-gathering and data-analysis skills.
   - Practice accurate observation skills.
   - Apply their results toward resolution of the original problem.
   - Use their enhanced understanding of the location under study to make predictions about similar problems.
   - Communicate their enhanced understanding of population biology to others.

Goal 4: To develop reasoning skills with application to science.

Students will be able to

1. State a purpose for all modes of communication, their own as well as those of others.
2. Define a problem, given ill-structured, complex, or technical information.
3. Formulate multiple perspectives (at least two) on a given issue.
4. State assumptions behind a line of reasoning.
5. Provide evidence and data to support a claim, issue, or thesis statement.
6. Make inferences, based on evidence.
7. Draw implications for action based on the available data.

Lesson Organizational Chart

This chart is a graphic depiction of the curriculum framework as it is incorporated within the lesson plans. Each lesson is listed under the heading of the primary type of goal that is covered (concept, content, or process). Secondary goals covered by the lesson plans are listed in parentheses.

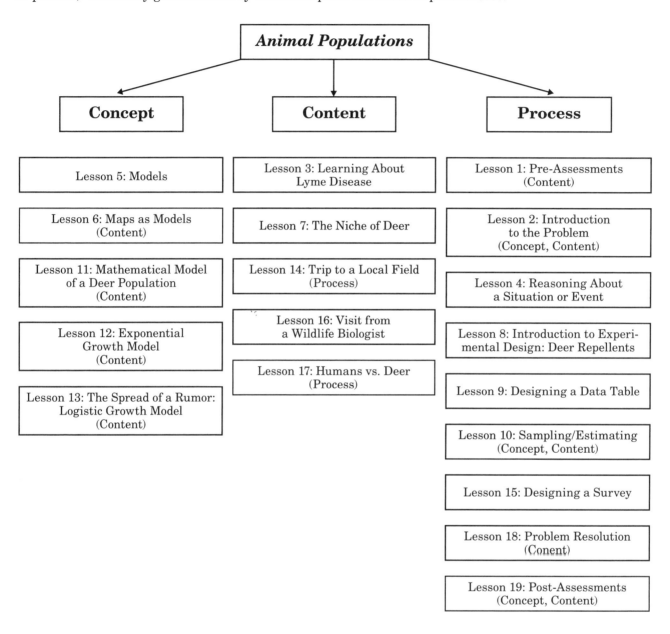

Content Background Information: Lyme Disease

There are several websites that offer extensive information about Lyme disease, its causes, and treatment:

American Lyme Disease Foundation: http://www.aldf.com/

Centers for Disease Control, Division of Vector-Borne Infectious Diseases: http://www.cdc.gov/ncidod/dvbid/lyme/index.htm

Lyme Disease Association, Inc.: http://www.lymediseaseassociation.org/

Medline Plus, a service of the U.S. Library of Medicine and the National Institutes of Health: http://www.nlm.nih.gov/medlineplus/lymedisease.html

Glossary

Assumption Conclusions based on one's beliefs and presuppositions

Bias A one-sided or slanted view that may be based on culture, experience, or other aspects of one's background

Biological Carrying Capacity The maximum population that the environment can support for an extended period of time without damage to the habitat

Biotic Potential The inherent capacity of an organism to increase in number; the maximum potential population growth rate

Buck A male deer

Community All of the organisms living in a given area

Conceptual Model A model created by making the conceptual leap of seeing similarities between something we do not understand (the thing we want to model) and something we understand well (the conceptual model). When we use analogies or metaphors to describe something, we are using conceptual models. For example, we could say that a muscle is like a rubber band, then think about the properties of rubber bands and use our understanding of these to make predictions about the properties of muscles. Note that while the rubber band is itself a physical thing, it is being used here as a conceptual model of a muscle.

Constant The factor(s) in an experiment that remain the same

Control The standard or baseline for comparing experimental effects

Cultural Carrying Capacity The maximum animal population that a human population is willing to tolerate in a given area

Dependent Variable The experimental variable that responds to changes in the independent variable

Doe A female deer

Ecology The study of ecosystems

Ecosystem A system composed of all living things (community) and the physical environment in the area

Edge The boundary between two different kinds of habitat (i.e. between forest and meadow). The edge can itself be a habitat.

Environmental Resistance the sum total of the environmental limiting factors acting on a population (predation, lack of food, etc.)

Exponential Growth A rate of growth characterized by rapid increases that get larger over time. Typically exponential growth functions involve a mathematical model of the form $y = (\text{constant})^x$, such as $y = 2^x$ (the variable is used as the exponent of a constant base).

Fawn A young deer, especially one still unweaned or retaining a distinctive baby coat

Habitat The kind of place or environment where a given organism normally lives

Hypothesis A tentative explanation for an observation, phenomenon, or scientific problem that can be tested by further investigation

Implication A suggestion of likely or logical consequence; a logical relationship between two linked propositions or statements

Independent Variable The variable that is changed intentionally by the experimenter

Inference Interpretation based on observation

Logistic Growth Model A model of population growth that acknowledges that the rate of population increase may be limited by environmental factors

Mathematical Model A mathematical relationship that behaves in the same way as the thing being modeled does. These can be simple (distance = rate × time) or complicated (a complex computer program that is used for weather prediction).

Mean The average value of a set of numbers

Median The middle value in a distribution of numbers

Mode The value or item occurring most frequently in a series of data

Model A structural design or miniature representation

Niche The role or position of an organism in the ecosystem, determined by its behavior and relationships to other ecosystem components. Note that a niche is not the same as a habitat.

Pathogen A microorganism that causes disease

Perspective An attitude, opinion, or position from which a person understands a situation or issue

Physical Model An actual physical device or process that behaves like the object or phenomenon being modeled, but represents a simplified version. (Example: a model rocket, when it is being used to study ballistic properties of real spacecraft: it looks and acts a lot like the real thing, but is smaller, costs a lot less, weighs less, cannot go as far, etc.)

Point of Inflection The point on a graph where the curve changes from concave down to concave up or vice versa

Point of View How people understand/look at things; what people think; the different ways people see things

Population Group of individuals belonging to the same species

Population Density The number of individuals per unit of area

Quadrat A small, usually rectangular plot of land arranged in a group for close study of plants or animals in an area

Range The difference between the smallest and largest values in a frequency distribution

Reasoning Evidence or arguments used in thinking

Scale The size of a representation in proportion to the size of the actual items, as on maps or models

Stakeholder An individual with an interest in or involvement with an issue and its potential outcomes

Survey (1) a detailed inspection of a geographic area for evidence of an organism; (2) a gathering of a sample of data or opinions considered to be representative of a whole

Transect A division of something such as a field, pond, or lagoon that is made by dividing the whole into strips. The term comes from the Latin for "cutting across."

Validity Accuracy and generalizability of the experiment's outcomes

Vector An organism that transmits a pathogen; causive agent of disease

The following is a comprehensive list of materials used during the course of this unit. The materials are also listed at the beginning of each lesson. In addition, handouts that accompany each lesson may need to be copied. As teachers tailor this unit for their own locations, they may decide to use materials other than those listed.

Animal Populations Materials List

Lessons	Items Needed
16	Audiovisual equipment for guest speaker
8	Bar of bath soap
14	Black construction paper (optional)
10	Box of 500 to 1000 styrofoam peanuts or lima beans
6	Buckets for waste
14	Camera and film
2, 16	Chart paper
8	Cheesecloth or nylon net
6	Clear plastic boxes with clear lids (such as shoe storage boxes)
8	Copies of school district policy on the use of animals in student research
12	Data regarding deer populations over a number of years in your locality or state
11, 12, 13	Graphing calculators
11, 12	Graph paper
6	Grease pencil for each group of 3 to 4 students
6, 15	Maps of local area (street, topographic, tax, orthophoto, etc.)
2, 16	Markers
14	Measuring tapes
17	*Noah's Garden: Restoring the Ecology of Our Own Backyards* by Sara B. Stein
6	Modeling clay (oil-based; not Play Doh)
14	Notepads (one per student)
14	Packets of salt, such as ones given at fast food restaurants (optional)
3	Reference materials about Lyme Disease
7	Resource materials about deer
6	Ruler for each group of 3 to 4 students
2	Storage system for information collected by students (3-ring binders; file folders; file boxes, etc.
14	Video camera and blank videotape
10	Videotape of shark estimation from *The Challenge of the Unknown* (optional)
6	Water pitchers and water

Handout Explanation

The handouts listed below are used throughout the unit to help structure and guide students' thinking about experimental design. A brief purpose statement is provided here about each handout.

Teachers can choose to post copies of these handouts around the classroom for students to reference while designing experiments instead of providing new copies in every experiment lesson.

Experimental Design Planner: *Forces students to think about the nuts and bolts of a good experiment; the intellectual details of the experiment*

Experimental Design Planner Checklist: *Provides students with standards for their experimental design and with the structure for revising it*

Experimental Protocol: *Provides students with a structure for the process of explicating the planned materials, methods, and data to be collected*

Laboratory Report: *Provides students with a list of important questions regarding the experiment they conducted and the results*

Problem Log: *The student's Problem Log is a written compilation of the student's thoughts about the problem. Most lessons contain suggested questions or a handout with questions for students to answer in their Problem Logs. The Problem Log should also be used by the student to record and store data and new information that they have obtained during the course of the unit. (For the Problem Log, each student should use either a three-ringer binder or a composition book.)*

Student Brainstorming Guide: *Asks students questions designed to guide their thinking about possible experiments to answer the problem*

Tailoring Animal Populations to Your Location

Classroom experience demonstrates that this unit is much more powerful when tailored for the location in which it is being presented. Accordingly, the following steps may assist your unit preparation.

1. Modify the problem statement so that the events occur at actual locations in your area.

2. Use local highway, topographic, and resource maps. This customization has several potential advantages: the site is familiar to the students; there are real, detailed topographic and street maps available; and the students interact with the actual community members who would be involved in managing deer populations.

3. Involve local experts (wildlife management personnel, extension office agents, biologists, government officials, etc.) as speakers and ongoing resources in the problem-solving process.

4. Work with media specialists to plan the unit and to assist students in finding resources. Other sources of support include special libraries (museums, corporations, historical societies, etc.) whose staffs offer vast knowledge of resources relevant to the unit.

5. Look for similar situations cited in the newspaper and use them as resources or points of comparison.

6. If it is not possible to tailor the unit to your location because there are no deer, you may use the deer data already included in the unit, or you may want to change the offending organism to one that is a problem in your area. Some suggestions are: Japanese beetles, starlings, wolves, rodents, mosquitoes, beaver, foxes, mice, bats, etc. You will need to make other major changes as the tick problem will not be associated with these other organisms.

7. The following are resources that might be useful in tailoring the unit:

   • A news article on "country critters" invading American towns: "Backyard Bears" in *Newsweek* (September 9, 1996, pp. 76–77).

   • Laurie Garrett's *The Coming Plague: Newly Emerging Diseases in a World Out of Balance.* (Farrar, Straus, & Giroux, 1994). The chapters on Lyme disease and hantavirus infection will be particularly useful.

Field Work Safety Precautions

Each school district has safety rules that should be strictly followed in the implementation of this unit. In addition, we suggest specific safety precautions relevant to the field portions of this unit:

1. Students will visit a deer habitat in the woods and the adjoining grassy areas. This is also a prime tick habitat. To avoid tick bites, students may wear insect repellent supplied by their parents. In addition, they should be encouraged to wear long sleeves and long pants, preferably light in color so that ticks crawling on clothing can be seen and removed. Sturdy shoes should be worn with socks.

2. If students are bitten by ticks, they should be instructed on procedures for safe removal. Note that tick removal is considered a medical procedure by some schools. Thus, teachers and school personnel may not be allowed to remove ticks from students' bodies.

3. Ticks are best removed with tweezers or by wrapping the tick in tissue paper and pulling it out with fingers. Do not twist or jerk, but pull slowly to avoid leaving the mouth parts in the wound. Do not use nail polish, petroleum jelly, alcohol, or hot matches to remove the tick. Wash the wound with an antiseptic after the tick is removed. Kill the tick in rubbing alcohol and keep it in a small vial for a few months in case any disease symptoms develop. *Source: Paul Davis, Extension Agent, New Kent County, Virginia.*

4. A recent study (Shih, C. M. and Spielman, A., "Topical prophylaxis for Lyme disease after tick bites in a rodent model," *Journal of Infectious Diseases* (IH3) 1993 Oct; 168 (4):1042–5.) reported that (in mice) transmission of Lyme disease from infected ticks can be substantially reduced by treating the tick bite with a topical antibiotic. We therefore recommend that tick bites be treated with Neosporin or some other over-the-counter antibiotic preparation.

5. Experimentation in this unit may involve the use of animals. Your district will have rules for animal experimentation; be sure to become familiar with these rules and follow them. Any investigations or experiments dealing with wildlife should be monitored to ensure protection for the organisms studied. Wildlife should not be harmed.

Student Signature   Date

Teacher Signature   Date

Dear Family,

Your child is about to begin a unique science experience that utilizes an instructional strategy called problem-based learning. In this unit, students will take an active role in identifying and resolving a real-world problem constructed to promote scientific learning. Your child will be gathering information from a variety of sources both in and out of school in order to contribute to the problem resolution. Goals for the unit are as follows:

Goal 1 To understand the concept of models

Goal 2 To use the interdisciplinary principles of population biology

Goal 3 To use scientific process skills to identify and solve problems

Goal 4 To develop reasoning skills with application to science

Good curriculum and instructional practice should involve families as well as teachers. We know from educational research that family involvement is a strong factor in promoting positive attitudes toward science, and we encourage you to extend your child's school learning through activities in the home.

Ways you may wish to help your child during the learning of this unit include:

- Discuss the progress of the unit and its issues with your child. This unit will provide many opportunities for family discussions at the dinner table or in the car.

- Allow your child to describe the problem and the day's outcomes to you, and try to solve the problem along with your child.

- Assist your child in finding community resources related to the problem.

- Engage your child in scientific experimentation exercises based on everyday events. For example, in a grocery store, how would you test whether it is better to go in a long line with people who have few items or a short line with people who have full carts?

- Visit area science museums and the library to explore how scientists solve problems.

- Use the problem-based learning model to question students about an issue they might have about the real world. For example: How does hail form? Answer: What do you know about hail? What do you need to know to answer the question? How do you find out?

Thank you in advance for your interest in your child's curriculum. Please do not hesitate to contact me for further information as the unit progresses.

Sincerely,

2

Lesson Plans

Pre-Assessments

1

Instructional Purpose

- To assess student understanding of experimental design
- To assess student understanding of population biology
- To assess student understanding of models

Curriculum Alignment    Goal 1 **Concept**    Goal 2 **Content**    Goal 3 **Process/ Experimental Design**   ◯ Goal 4 **Process/ Reasoning**

 ## Materials/Resources

- Experimental Design Pre-Assessment (Handout 1.1)
- Experimental Design Rubric (Teacher Resource 1)
- Content/Concept Pre-Assessment (Handout 1.2)
- Content/Concept Pre-Assessment Scoring Guide (Teacher Resource 2)
- Process Pre-Assessment (Handout 1.3)
- Process Pre-Assessment Scoring Guide (Teacher Resource 3)

Lesson Length

90 minutes

 ## Activities

1. Explain to students that they will be beginning a new unit of study focused on scientific inquiry. Tell students that in order to get a sense of how much they already know and to be able to tell how much they have learned by the end of the unit, they will need to take several pre-assessments.

2. Distribute the **Experimental Design Pre-Assessment** (Handout 1.1) and have students complete it individually. Collect and score the assessments using the **Experimental Design Rubric** (Teacher Resource 1).

3. Distribute the **Content/Concept Pre-Assessment** (Handout 1.2) and have students complete it individually. Score **Content/Concept Pre-Assessment** using **Content/Concept Pre-Assessment Scoring Guide** (Teacher Resource 2).

4. Distribute the **Process Pre-Assessment** (Handout 1.3) and have students complete it individually. Collect and score the assessments using the **Process Pre-Assessment Scoring Guide** (Teacher Resource 3).

5. Have students discuss which aspects of the pre-assessment they found difficult. Explain that throughout the unit they will be thinking about challenging questions that relate to concepts on the pre-assessment.

 ## Problem Log

Have students respond to one of the following prompts:

- When I want to know more about a topic in science, I . . .
- Studying science can help me in the following ways . . .
- It is important to know about science because . . .

 ## Notes

1. Send home the parent letter with each student who will be participating in the unit.

2. The pre-assessment given in this lesson serves multiple purposes. Performance on the pre-assessment should establish a baseline against which performance on the post-assessment may be compared. In addition, teachers may use information obtained from pre-assessments to aid instructional planning, as strengths and areas for improvement among students become apparent.

3. Students should have a unit notebook and folder that they can use throughout the unit to respond to Problem Log questions, and keep other written assignments and handouts from the unit. The notebook can also hold a running list of unit vocabulary, which also can be displayed in the classroom in chart form.

 ## Homework

Have students write a list describing attributes of scientists. Encourage them to generate as many different ideas as they can.

Assessment

- Pre-Assessments
- Problem Log

Technology Integration

- If the resources are available, students may keep their Problem Logs in electronic format rather than in paper and pencil format.

Name _____   Date _____

Experimental Design Pre-Assessment (30 minutes)

Construct a fair test of the following question: *Which brand of paper towel (Brand A, Brand B, or Brand C) holds the most water?*

Describe in detail how you would test this question. Be as scientific as you can as you write about your test. Write the steps you would take to find out which of the three brands holds the most water.

Adapted from Fowler, M, (1990). The diet cola test. *Science Scope, 13(4)*, 32–34.

Teacher Resource 1: Experimental Design Rubric

Criteria	Strong Evidence 2	Some Evidence 1	No Evidence 0	Pre	Post
States **PROBLEM** or **QUESTION**.	Clearly states the problem or question to be addressed.	Somewhat states the problem or question to be addressed.	Does not state the problem or question to be addressed.		
Generates a **PREDICTION** and/or **HYPOTHESIS**.	Clearly generates a prediction or hypothesis appropriate to the experiment.	Somewhat generates a prediction or hypothesis appropriate to the experiment.	Does not generate a prediction or hypothesis.		
Lists experiment steps.	Clearly & concisely lists four or more steps as appropriate for the experiment design.	Clearly & concisely lists one to three steps as appropriate for the experiment design.	Does not generate experiment steps.		
Arranges steps in **SEQUENTIAL** order.	Lists experiment steps in sequential order.	Generally lists experiment steps in sequential order.	Does not list experiment steps in a logical order.		
Lists **MATERIALS** needed.	Provides an inclusive and appropriate list of materials.	Provides a partial list of materials needed.	Does not provide a list of materials needed.		
Plans to **REPEAT TESTING** and tells reason.	Clearly states a plan to conduct multiple trials, providing reasoning.	Clearly states a plan to conduct multiple trials.	Does not state plan or reason to repeat testing.		
DEFINES the terms of the experiment.	Correctly defines all relevant terms of the experiment.	Correctly defines some of the relevant terms of the experiment.	Does not define terms, or defines terms incorrectly.		
Plans to **MEASURE.**	Clearly identifies plan to measure data.	Provides some evidence of planning to measure data.	Does not identify plan to measure data.		
Plans **DATA COLLECTION.**	Clearly states plan for data collection, including note-taking, the creation of graphs or tables, etc.	States a partial plan for data collection.	Does not identify a plan for data collection.		
States plan for **INTERPRETING DATA.**	Clearly states plan for interpreting data by comparing data, looking for patterns and reviewing previously known information.	States a partial plan for interpreting data.	Does not state plan for interpreting data.		
States plan for drawing a **CONCLUSION BASED ON DATA.**	Clearly states plan for drawing conclusions based on data.	States a partial plan for drawing conclusions based on data.	Does not state plan for drawing conclusions.		

TOTAL SCORE:

Adapted from Fowler, M. (1990). The diet cola test. *Science Scope, 13(4)*, 32–34.

Name   Date

Content/Concept Pre-Assessment (30 minutes)

1. Describe the role of deer in the spread of Lyme disease.

2. List three possible ways to control deer populations. For each method, list one advantage and one disadvantage.

3. Give an example of each:

   a. a physical model

   b. a conceptual model

   c. a mathematical model

Content/Concept Pre-Assessment Scoring Guide (Teacher Resource 2)

1. Describe the role of deer in the spread of Lyme disease.

   Answer: White-tailed deer are parasitized by deer ticks. Deer ticks are frequently infected with Borrelia burdorfi, *the organism that causes Lyme disease. When an infected tick bites an uninfected deer, the deer can become infected. Uninfected ticks that bite the infected deer then become infected themselves. Because deer can travel great distances, infected ticks can be spread into new territory. When infected ticks drop off the deer, they can then move onto humans or household pets. A bite from an infected tick will cause a person to become infected and to develop Lyme disease. Deer thus have two major roles in the spread of the disease: they facilitate the infection of previously uninfected ticks, and they allow infected ticks to move from place to place.*

   (Accept any answer that mentions the two roles described above.)

2. List three possible ways to control deer populations. For each method, list one advantage and one disadvantage.

   Accept any reasonable answers. Here are two examples.

   * *Hunting: Advantage is efficacy; disadvantage is public attitude.*

   * *Starvation and disease: Advantage is that it is natural; disadvantage is public reaction and effects on the ecosystem of desperate attempts of deer to find food.*

3. Give an example of each: (accept all reasonable answers)

   a. a physical model—*scale model of a building*

   b. a conceptual model—*the idea of a light bulb could be a model for a brain getting an idea; it lights up when the idea occurs in the brain.*

   c. a mathematical model—*an equation such as $D = R \times T$*

Name _____   Date _____

Process Pre-Assessment (30 minutes)

In a *New York Times* article ("Roissy Journal: Invincible Rabbit Army Besieges a Paris Airport," Thursday, August 8, 1996, page A4), the journalist, Craig Whitney, described a problem faced by the managers of Charles de Gaulle Airport. The airport is overrun with rabbits. The gamekeeper Jean Valissant estimates that there are 50,000 of them living in the 5,000 acres of fenced-in grass and short cornfields that line the two airport runways.

1. One of Jean Valissant's tasks is to estimate the number of rabbits present at the airport.

    a. If you were the gamekeeper at the airport, what equipment and resources would you need to make your estimate?

    b. What procedure would you use to make your estimate? Describe what you would do and how you would use the results to estimate the number of adult rabbits present at the airport.

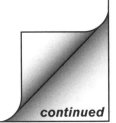

continued

c. How could you check to be sure that your estimate was accurate? Describe the procedure you would use and how you would use your results to check your estimate.

d. List at least two models that you would use during your estimation procedure and explain whether they are physical models, conceptual models, or mathematical models. What limitations does each model have?

Process Pre-Assessment Scoring Guide
(Teacher Resource 3)

In a recent *New York Times* article ("Roissy Journal: Invincible Rabbit Army Besieges a Paris Airport," Thursday, August 8, 1996, page A4), the journalist, Craig Whitney, described a problem faced by the managers of Charles de Gaulle Airport. The airport is overrun with rabbits. The gamekeeper Jean Valissant estimates that there are 50,000 of them living in the 5,000 acres of fenced-in grass and short cornfields that line the two airport runways.

1. One of Jean Valissant's tasks is to estimate the number of adult rabbits present at the airport.

   a. If you were the gamekeeper at the airport, what equipment and resources would you need to make your estimate?

   You would need nets or traps to catch them, tags or some other way of marking rabbits that had been caught, protective gear (such as gloves) to protect people handling these wild animals, some way to manage and analyze the data (everything from paper and pencils to a computer) and people to do the work.

   (Accept any reasonable answer.)

   b. What procedure would you use to make your estimate? Describe what you would do and how you would use the results to estimate the number of adult rabbits present at the airport.

   I would use the tag and recapture method to estimate the rabbit population. On day one, my helpers and I would set humane traps loaded with rabbit bait at random points around the airport. We would examine the traps the next day, tag any rabbits we found, and set them free. A few days later, we should set humane traps loaded with rabbit bait in a new set of random points around the airport (to avoid biasing our results in favor of rabbits who were in the habit of hopping past the previous trap sites.) We would examine the traps the next day and determine what fraction of the rabbits were tagged. We would then use proportional reasoning to estimate the total size of the rabbit population.

   (Accept any reasonable answer.)

   c. How could you check to be sure that your estimate was accurate? Describe the procedure you would use and how you would use your results to check your estimate.

   My helpers and I would capture, tag, and count all of the adult rabbits in several small areas of the airport and determine the average number of rabbits per unit area. We would then multiply this number by

continued

the total area of the airport to estimate the total number of adult rabbits at the airport. If this estimate and the tag and recapture estimate were similar, we could feel confident that we had a good estimate of the number of adult rabbits at the airport.

(Accept any reasonable answer as long as it uses a different method of estimation than the one mentioned in the student's answer to part b.)

d. List at least two models that you would use during your estimation procedure and explain whether they are physical models, conceptual models, or mathematical models. What limitations does each model have?

- *I would need a map of the airport to choose sites for traps and for intensive rabbit counting. This is a physical model, depending on the kind of map I was using, different kinds of information would be missing (there's no way to tell from an aerial photo whether the are snakes in the grass where I'm planning to work, for example) so a trip to the potential trapping sites would be necessary to confirm their usefulness.*

- *My estimates would be based on mathematical equations, which are mathematical models. Which equations to use is dependent on my assumptions about the rabbits and the airport (for example, in the tag and recapture method I am assuming that rabbits that had been caught once would be equally likely to be caught a second time and that the tagged rabbits will be evenly distributed across the airport when it's time to recapture them). These assumptions may or may not be valid; the validity of the assumptions limits the usefulness of the model.*

- *Accept any other reasonable answers.*

Introduction to the Problem

2

Instructional Purpose

- To introduce the problem
- To allow students to begin to formulate a description of the problem
- To catalyze student-directed research in the problem area

Curriculum Alignment   Goal 1 Concept   Goal 2 Content   Goal 3 Process/ Experimental Design   Goal 4 Process/ Reasoning

 **Vocabulary**

Stakeholder An individual involved in the resolution of a problem-based learning problem.

Assumption Conclusion based on one's own beliefs and presuppositions

Materials/Resources

- Chart paper
- Markers
- Newspaper article (copy for each student; see note at end of lesson)
- Storage system for information collected by students: three-ring binders, file folders, file boxes, etc.

- Problem Statement (Handout 2.1)
- Need to Know Board (Handout 2.2)
- Resource Summary Form (Handout 2.3)
- Problem Log Questions (Handout 2.4)

Lesson Length

Three 60-minute sessions; may be done in two 90-minute sessions (need to give students time to look for relevant information)

Activities: The Problem Statement

1. Assemble materials and handouts; tape three large sheets of chart paper to the wall

or chalkboard at front of room. Label the leftmost sheet "What Do We Know?", the center sheet "What Do We Need to Know?", and the rightmost sheet "How Can We Find Out?"

2. Distribute copies of the **Problem Statement** (Handout 2.1); ask the students to place the problem statement in their **Problem Logs.** (For the **Problem Log,** a three-ring binder works well for maintaining all of the handouts associated with the unit. A composition book may also be used; students will need to tape or glue the handouts in it.) Then ask the students to read the statement carefully.

3. Ask the students to tell you what they know based on the information in the problem statement; record this information on the "What Do We Know?" portion of the **Need to Know Board**. Gently challenge assumptions; require students to justify each item you record. Questions such as "How do we know this?" and "Are we making an assumption here?" will facilitate this process. Have students record the information on their own **Need to Know Board** (Handout 2.3).

4. Once the students have stated what they know, move to the "What do we need to know?" portion of the **Need to Know Board.** Ask students to tell you what they need to know based on the information in the problem statement. Require them to justify their answers. Questions such as "Why do we need to know this?" and "Why is it important to find out about this?" will facilitate this process. At this stage, it will also be useful to ask students about their understanding of the situation faced by the stakeholder in the problem (Marie Barnes). Questions such as "What is going on here?" "What are we supposed to do?" and "Do we have a problem?" will help the students to focus their questioning. Ask the students to prioritize their list of questions based on their understanding of the problem situation.

5. Move to the "How can we find out?" section of the **Need to Know Board.** For each question on the "What do we need to know?" section of the board, ask students to suggest ways that the answer could be found. Some of the questions will require you to give the students additional information about the stakeholder and his problem; others will require the students to use available information resources. Assign questions that do not require your intervention to small groups of students; come to an agreement about the deadline for their having retrieved the information. Ask students to write their assigned question and the deadline for answering it in their **Problem Logs.** Remind them that any information they collect will need to be written in their **Problem Logs** as well. Students will also write a summary of information they find and will file it in a class storage system.

Activities: Consulting with a Media Specialist

1. Before this session, arrange for your media specialist to be available to speak with the students. Have students go to the media center with needs already in hand; have the media specialist facilitate information searches using books, periodicals, the Internet, microfiche, CD ROM database, etc. (Remind students to evaluate the reliability of each resource. This could be handled as a separate lesson by the media specialist or a language arts teacher.)

2. Prepare the media specialist in advance with questions from the **Need to Know Board** so that he/she knows in advance what the information needs are likely to be. Have the media specialist suggest other sites outside of school to get information (e.g., university library, regional library, online possibilities, local agencies, people resources).

 ## Activities: Summarizing Relevant Information About the Problem

1. When the students arrive in class, ask them what information they have acquired for the **Need to Know Board.** Record the new information and include the source.

2. Tell the students that they will be expected to write a short summary of each information resource they find. To demonstrate, either find a newspaper article on Lyme disease (not an Internet or book source; see note to teacher at the end of this lesson for suggestions), or collect a suitable resource from a student. Allow the students to work individually to summarize the article using the **Resource Summary Form** (Handout 2.3). Then debrief about it as a group.

3. As a class, discuss the best way to manage the information that will be collected about the problem. Remind the group that everyone in the class will need to have access to all of the information, not just to the data they collected themselves. Talk to them about organizing the summaries. An information storage system could take the form of a bulletin board, file folders, three-ring binders with transparent sheet holders, an index card box, or an electronic database. Either design your own storage system and show it to your students or have them design one as a class. Make a list of keywords for filing the summaries; the list can be added to if necessary. If you are using an index card box, each summary should be filed under one keyword (the most descriptive available); people using a computer database could cross-file the summary under several key words.

4. Ask students to write summaries of the information they have gathered, and file them using the class system.

 **Notes**

1. This is the profile of our stakeholder. Share information with students as needed.

   Marie Barnes is the vice-mayor of Kingsford, a small town on the East Coast of the U.S. Her responsibilities include community outreach and troubleshooting for the mayor. Her expertise is urban planning. She has a big conference over the Memorial Day weekend. Her husband, Chris, is a wildlife biologist who has just started his new job in the local office of the state fish and wildlife service. They have three children, two boys (Sam, age 11 and Josh, age 9) and a girl (Patricia, age 5). The children are in daycare with a neighborhood sitter when they are not in school. The family has just moved to the area (three weeks ago) from Los Angeles.

   E-mail address for Chris: E-mail address for Marie:
   crbarn@wlm.gov   mtbarn@etown.com

2. Please look ahead to the materials section of Lesson 10 for information about ordering *The Challenge of the Unknown* materials if you do not already have them. If *Challenge of the Unknown* materials are unavailable, the DVD *Jaws of the Pacific* from the Discovery Channel website will serve the same purpose.

3. Any major newspaper, such as *The Washington Post, The Wall Street Journal,* or *The New York Times* should have recent articles on Lyme disease in their archives. Local newspapers may have such articles as well.

4. The word "research" is used in this unit to mean either library research or scientific research. The difference between the two kinds of research is that library research is secondhand data gathering (collecting information found by others) and scientific research is firsthand data gathering (investigating a problem through experimentation).

 Homework

Assign the **Problem Log Questions** (Handout 2.4) as homework. Have the students answer these questions in their **Problem Logs.**

 Assessment

- Observation of individual students during class discussion. (A checklist allowing you to record the participation of individual students may be helpful here.)
- Answer to **Problem Log Questions.** (Do students seem to grasp the fundamentals of the situation? Can they justify their interpretations? If not, you may want to return to the **Need to Know Board** and discuss it in more detail.)

Technology Integration

- For the information storage system, students may want to use an electronic database.
- If the resources are available, students may keep their **Problem Logs** in electronic format rather than in paper and pencil format.

Problem Statement: E-mail from Marie Barnes

	Message	Insert Options Write

```
Send   Send   Font  ▼ 12 ▼   Attach
              B  I  U ▼ A ▼   File
```

To	
Cc	
Subject	

From: Marie Barnes <mtbarn@etown.com>
Subject: Dinner
To: Chris Barnes <crbarn@wlm.gov>
Date: Friday, 25 May 2005

Chris,

I'm not going to be able to make it home tonight after all—sorry! The evening conference session will be much more interesting than I had thought and so I'm going to be staying at Sue's house tonight rather than coming all the way home. Could you please pick up some Chinese food on your way home and get the kids from daycare? Oh, and the babysitter left a message this morning that Josh has a funny rash on his tummy. I hope it isn't serious—Dr. Martin's office will be closed until after Memorial Day. Could you take a look at it and see what you think?

Love,
Marie

P. S.
Here's what the babysitter said:
Josh has a large, red swelling right next to his navel. He says that he had what looked like a bug bite there last week; since it didn't itch or hurt, he didn't worry about it. Yesterday night he thought it looked bigger; now it's almost an inch in diameter. It's really weird-looking: it looks like a red ring with a white center. It looks as though there was a bug bite in the very center of the swelling. It still doesn't itch or hurt, but he's pretty upset about it.

Name _____   Date _____

Need to Know Board

What we know . . .	What we need to know . . .	How we can find out . . .

Handout 2.3

Resource Summary Form

Name _____ Date _____

Resource Summary Form

Name of the person finding the resource: _____

Resource citation: Author or editor, title, date (copyright or periodical), page numbers, publisher, place of publication.

Summary:

What information did you find that answers a question on the Need to Know Board?

42

Copyright © Kendall/Hunt Publishing Company

Name _____   Date _____

Problem Log Questions

1. What do you think is the most important statement on the "What Do We Need to Know" list?

2. Explain why you think this statement is so important.

3. Which statement needs to be answered first? Explain your answer.

Learning About Lyme Disease

Instructional Purpose

- To engage students in research about the topic of Lyme disease

Curriculum Alignment ○ Goal 1
 Concept ● Goal 2
 Content ○ Goal 3
 Process/
 Experimental Design ○ Goal 4
 Process/
 Reasoning

 ## Vocabulary

Pathogen A microorganism that causes disease

Vector An organism that transmits a pathogen/causive agent of disease

 ## Materials/Resources

- Reference materials about Lyme disease
- E-mail from Chris (Handout 3.1)
- Memo from the Mayor/Letter from *The Kingsford Gazette* (Handout 3.2)
- Problem Log Questions (Handout 3.3)

Lesson Length

60 minutes

 ## Activities

1. Ask the students to report any information they have gathered since the last class session; discuss as needed. Update the Need to Know Board.

2. Give the students the **E-mail from Chris** (Handout 3.1). Ask them to update the Need to Know Board based on information gleaned from the e-mail.

3. Give students **Memo from the Mayor/Letter from *The Kingsford Gazette*** (Handout 3.2). Ask them to update the Need to Know Board based on its contents.

4. Students should find their own information about Lyme disease when possible. Remind students that this problem solving experience means

an open-ended search for information; they are not looking for a preset answer or set of data. Also remind them that the Internet is not the only place to get information and that the Internet is not always a reliable source. They should also use sources such as newspapers, library databases, and books that you have put on reserve at the library.

Assign information-gathering tasks related to Lyme disease. Set deadlines for each task. Topics that should be covered include:

   a. What causes it?

   b. What are its effects on the human body?

   c. How is it transmitted?

   d. How can its transmission to humans be prevented?

   e. How is it treated?

   f. What is the relationship of deer to Lyme disease?

   g. What other diseases are passed in the same fashion?

5. Modify the class data collection system as necessary and as the problem expands/changes.

6. Monitor the progress of student research by meeting frequently with the small groups and by reading student **Problem Logs.**

Notes

1. A good reference book for students to use during this lesson is *Your Child's Health: The Parent's Guide to Symptoms, Emergencies, Common Illnesses, Behavior, and School Problems* (revised edition), by Barton Schmitt, MSD, FACP. It is published by Bantam Books (2005). ISBN 0553383698

2. It is a common misconception that deer are the only host for the deer tick (vector for Lyme disease). A more important host on the east coast of the United States is the white-footed mouse. The following article will provide good background information:

   Ostfeld, R. S., Jones, C. G., & Wolff, J. O. (May 1996). Of mice and mast: Ecological connections in eastern deciduous forests. *BioScience, 46* (5), pp. 323–330.

3. Recommended teacher background reading is:

   "The Ecology of Lyme-Disease Risk," by Richard S. Ostfeld in the *American Scientist,* Volume 85 (July/August, 1997), pp. 338–346.

Homework

Have the students complete the research assignments as homework.

Extending Student Learning

Provide the following options for students who would like to extend their learning:

- Research trends in the number of Lyme disease cases in various states. **Ask:** *How has this changed over time? What are the implications for the future?*

- Visit the Centers for Disease Control online and get up-to-date data on Lyme disease. (http://www.cdc.gov) The electronic journals *Journal of Emerging Diseases* and *Morbidity and Mortality Weekly Report* (MMWR) are published by the CDC and have up-to-date information on disease patterns. Using information you find in these sources, develop visual materials about Lyme disease to display in the classroom.

- Web MD and similar sites will have information on Lyme disease. (http://www.webmd.com) Prepare a presentation for your classmates regarding the symptoms and treatment of Lyme disease.

 ## Assessment

- Completed **Need to Know Board**
- Active and appropriate participation in group and class discussions
- **Problem Log Questions:** Students should utilize information from the class discussion to frame their answers.
- **Problem Logs:** Monitor the progress of student research.

 ## Technology Integration

- Through the Internet, students may access the local, state, and national agencies that deal with wildlife issues. They may use this information in their ongoing research about the problem.

- This unit would be a good opportunity to utilize an E-mentor, a professional who is willing to correspond with students electronically regarding a certain body of knowledge. A physician, someone from a local wildlife agency, or someone from the health department could work with students to supplement their database and to provide guidance with the problem situation.

E-mail from Chris

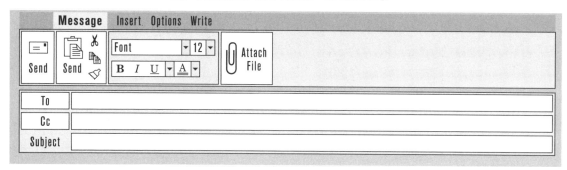

From: Chris Barnes <crbarn@wlm.gov>
Subject: Josh
To: Marie Barnes <mtbarn@etown.com>
Date: Saturday, 26 May 2005

Marie,

The doctor said that Josh has the classic Lyme disease bullseye rash—he says that nothing else looks like this. Josh is taking antibiotics for it now. The doctor thinks that we've caught it early enough.

Did you know that there are two other kids on our block who have come down with Lyme disease this spring? And Mrs. Spenser across the street says that she's had it twice since she moved here ten years ago. I guess that all the time she puts in her garden must expose her to a lot of ticks.

I had no idea that we lived in such an active place!

Chris

Memo from the Mayor/Letter from
The Kingsford Gazette

MEMO

To: Deputy Mayor Marie Barnes

From: Jane White, Mayor

Marie,

Did you see this one? I'd like something solid on controlling Lyme disease for the City Council meeting on Tuesday—I can't stall on this one for very long. I've had three phone calls about the Lyme "epidemic" this morning already.

Jane

May 26, 2005

From the Editor's Desk

Virginia Olin, Ed.

Today's lead article in the *Gazette* makes clear the magnitude of the local Lyme disease epidemic. Twenty cases in the last month alone—it staggers the imagination. Local wildlife experts blame the mild winter for a bumper crop of deer ticks; it is our firm opinion that the huge local surplus of deer is also responsible. It is time that city council solved this problem once and for all.

Name _____   Date _____

Problem Log Questions

1. How serious is Lyme disease for the individual who contracts it? How is it treated, if caught early? How is it treated if caught later?

2. What is the role of deer in the spread of Lyme disease?

continued

Name _____   Date _____

3. What is the role of mice in the spread of Lyme disease?

4. What do you think the main problem is now?

Reasoning About a Situation or Event

4

Instructional Purpose

• To develop reasoning skills with application to science

Curriculum Alignment Goal 1 Concept Goal 2 Content Goal 3 Process/ Experimental Design Goal 4 Process/ Reasoning

 Vocabulary

Assumption Conclusions based on one's beliefs and presuppositions

Bias A one-sided or slanted view that may be based on culture, experience, or other aspects of one's background

Implication A suggestion of likely or logical consequence; a logical relationship between two linked propositions or statements

Inference Interpretation based on observation

Perspective An attitude, opinion, or position from which a person understands a situation or issue

Point of View How people understand/look at things; what people think; the different ways people see things

Reasoning Evidence or arguments used in thinking

Stakeholder An individual with an interest in or involvement with an issue and its potential outcomes

 Materials/Resources

• Reasoning Wheel (Handout 4.1)

• Reasoning Wheel with Sample Questions for Article Discussion (Teacher Resource 1)

• Deer Population Article (Handout 4.2)

• Reasoning About a Situation or Event (Handout 4.3)

Lesson Length

120 minutes

 **Activities**

1. Explain that in this unit the class will be learning and using a certain way of thinking or reasoning. This method of thinking is something you can use in any situation, both in school and in your daily lives outside of school, and maybe even use it when talking to your parents or friends.

2. Distribute the **Reasoning Wheel** to the students (Handout 4.1) and have a copy on the board or on an overhead projector. Keep a copy of this wheel posted prominently in the classroom for the duration of the unit.

3. Explain to students that this way of thinking takes a while to learn and it is not going to be easy the first time. Today they are not trying to master the method, they are just getting introduced to it and they will return to it throughout the unit. Also remind students that this is just one way to help them in their thinking.

4. Tell students that they are going to discuss a newspaper article about an issue related to deer populations. Distribute **Deer Population Article** (Handout 4.2) and have students read it. Then tell students that they are going to analyze this selection as you introduce them to the reasoning method.

5. Work through the Reasoning Web in a question-answer format. Use **Reasoning Wheel with Sample Questions for Article Discussion** (Teacher Resource 1) for guidance. Write the class answers in the wheel and have students write the answers down as well as any notes about the reasoning method.

   Point to the **Purpose** box on the wheel. Tell students that the first step in reasoning is to decide what your purpose is. Any time we reason or think about something we have a goal for doing it, something we hope will be achieved or decided by our thinking, otherwise we wouldn't waste the time to do it. **Ask:**

   • What was the author's purpose for writing this piece?

   • Brainstorm answers, have the class discuss them, and develop a purpose statement they agree on to write in the box.

   Direct students to the Issue/Problem circle. Explain that when thinking about something we need to start with a clear question. Knowing exactly what we are looking for helps us to focus on important information and find an answer. **Ask:**

   • What is the larger issue or question the author addressing?

   • What does the author see as a problem he is trying to address?

   • Formulate a question as a class and write it in the circle.

Tell students that once you have identified the question you need to think about the point of view. **Ask:**

- What do we mean when we say point of view? *(How people understand / look at things, what people think, the different ways people see things)*

When we think about something we need to realize that there may be different ways of looking at it than our way of seeing the problem. **Ask:**

- How could it help our thinking to look at the problem from several other people's points of view? For example: One day you accidentally bump into your friend in the hallway and your friend gets really angry with you. You are thinking, "How rude, it was an accident," and you get upset. Your friend, on the other hand, just got in a fight with a boyfriend/girlfriend or got a bad grade on a test and was in a really bad mood. Your friend just blew up but it wasn't because of what you did. *(We see things we may miss, we may realize that we don't get the problem, we may be biased, or we may learn something from another point of view, the problem may seem different / less important / more important / important for different reasons from someone else's view.)*

By being aware of our own point of view and trying to see other's points of view, our thinking hopefully is less biased and our arguments stronger because we have considered the way others will view the problem. **Ask:**

- What was the author's point of view?

- What are some of the other points of view that should be considered? Who is he trying to get to listen to him? Write the people and their points of view in the web. Remind students that they need to keep all of these in mind as they take the next steps in reasoning.

Next, we need to look at the evidence. **Ask:**

- What is evidence and why is it important in answering any question? *(It is proof / information / facts that tell you something about the question, you need it to make sure you decide the right thing, you need it to convince other people to agree with your answer).*

Tell students that in reasoning we need to make sure our evidence is accurate and we need to look at evidence that supports and opposes our own ideas so that our answer is not biased. **Ask:**

- What evidence does the author have or use?

- Is it easy to identify his evidence?

- What evidence is missing?

- What other data would help his argument?

If students give assumptions, ask them what proof they have. Write them down and say you will come back to them. Tell students the next part of reasoning is to identify assumptions. **Ask:**

• What are assumptions? *(Things you think but don't have proof for, things we believe).*

• What happens if our assumptions are weak or inappropriate and we make a decision based on them? *(We get the wrong answers, we might make a bad decision).*

So it is important to realize what our assumptions are and to make sure they do not lead us to a wrong answer. **Ask:**

• What were some of the author's assumptions?

• How accurate do you think these assumptions are?

• How are assumptions related to point of view?

Tell students that once they have gathered data and thought about the points of view and their assumptions they are ready for making inferences. **Ask:**

• Does anyone know what "inference" means or what it means to infer something?

Tell students that "inference" means we look at the evidence, and, keeping in mind our assumptions, we conclude something from that data. For example, if we saw two crunched cars on the side of the road right next to each other and a policeman talking to people who were pointing at the cars, we would infer that those two cars had hit one another. Inferences are small steps we take in our mind about the evidence we have to help answer our question. These are tentative ideas, just suggestions to think about as possible answers. But there are degrees of inferences; some are small thoughts, while other times we have strong reactions to our inferences. **Ask:**

• What inferences does the author make?

The final step in thinking is to consider the consequences or implications of our inferences or of following through on a given point of view. **Ask:**

• What are implications? *(What or who will be affected by your decision, what your decision means for other people).*

• What are consequences? *(What will happen if you do something, what are the results of your actions).*

Before we make a decision we should think about who will be affected by it and what will be the results of our decision, and then think about what we should do. **Ask:**

• What are the implications of what the author is saying?

• What would happen if this grid system broke down across the entire United States?

• How do you think other people would respond to this?

6. Explain to the students that every time we reason about something we do not have to focus specifically on all these steps, because the emphasis may need to be just on some of them. Also, this is just one way to help us focus our thinking and make decisions. Tell them as they work through the unit this will become easier and it will help them as they study electricity and the importance of electrical systems for consumers.

7. To review the steps, have students work through them in small groups considering the **Problem Statement** (Handout 2.1) to complete a **Reasoning Wheel.** Have each group take the point of view of a different stakeholder group in this situation. (This could be done as homework or in class depending on the available time.) Have students share their findings and discuss as a class. Distribute **Reasoning About a Situation or Event** (Handout 4.3) for use in working specifically with stakeholder points of view, assumptions, and implications.

Note

The Reasoning Model introduced in this lesson, which is based on the work of Paul (1992), may be used in a variety of ways throughout the unit. The Reasoning Wheel can be used to consider issues and documents, and questions related to the Elements of Reasoning may be included in most lessons. You may wish to introduce the elements with reference to a current event in your school or community as a way of providing support for student understanding. In addition, using the vocabulary of the reasoning model on a regular basis can help students to learn to use the elements more effectively in considering issues. Some additional information on the Reasoning Model appears in the implementation section at the back of the unit.

Extending Student Learning

Have students find another article related to some issue concerning deer populations or Lyme disease. Have them complete **Reasoning About a Situation or Event** (Handout 4.3) in relation to the article.

Assessment

- Completed Reasoning Wheels: Evaluate whether students demonstrated an understanding of the terminology in relationship to the issue being discussed.
- Completed Reasoning About a Situation or Event: Evaluate whether students demonstrated an awareness of stakeholder points of view, assumptions, and implications.
- Class discussion participation

Technology Integration

- Students may use the Internet to research the issue of deer population problems.
- Students may use spreadsheet software to develop a chart or graph showing the instances of Lyme disease outbreaks in their localities.

Name _____   Date _____

Reasoning Wheel

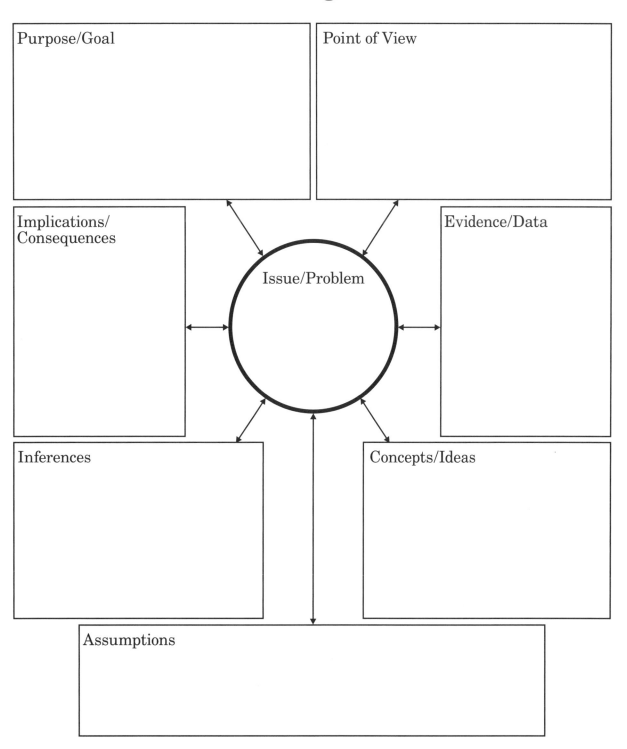

Purpose/Goal

Point of View

Implications/
Consequences

Evidence/Data

Issue/Problem

Inferences

Concepts/Ideas

Assumptions

Adapted from Paul, R. (1992). *Critical thinking: What every person needs to survive in a rapidly changing world*. Sonoma, CA: Foundation for Critical Thinking.

Reasoning Wheel with Sample
Questions for Article Discussion
(Teacher Resource 1)

Purpose/Goal

What was the author's goal in writing this article?

Do you think he or she meets this goal? Why or why not?

Point of View

What points of view are presented in this article?

What biases do you find in these points of view?

Implications/Consequences

What does the author see as the implications of this article?

What do you see as the implications of this article? Why?

Issue/Problem

What issue or problem is this article addressing?

Evidence/Data

What data or other pieces of evidence does the author provide?

How do these support his or her point of view?

How do you know these are reliable?

Inferences

What inferences does the author draw in this article?

What inferences can you draw from the evidence presented in this article?

Concepts/Ideas

What are some of the larger concepts and ideas the article covers?

Assumptions

What assumptions does the author make while writing the article?

Adapted from Paul, R. (1992). *Critical thinking: What every person needs to survive in a rapidly changing world.* Sonoma, CA: Foundation for Critical Thinking.

Deer Population Article

HelenaIR.Com Independent Record
September 22, 2006
http://www.helenair.com/articles/2006/09/22/helena/a01092206_04.txt

Task force ponders city with unchecked deer population

By LARRY KLINE - IR Staff Writer - 9/22/06

What would happen if city officials chose to do nothing to corral the growth of Helena's urban deer herd?

Members of the Urban Wildlife Task Force on Thursday considered that question as part of their analysis of lethal and non-lethal options the city might employ to control the deer population.

The group identified one merit in maintaining the status quo — the sight of deer in town is pleasing, members said—and plenty of potential problems.

Outlining issues associated with an unchecked deer herd roaming city streets and backyards allows the group to give the Helena City Commission and the public a clearer picture of possible strategies, state Fish, Wildlife and Parks biologist Gayle Joslin said.

Authorities already manage deer in some ways. FWP wardens and city animal control personnel euthanize injured deer if they cannot walk and move on their own. A city ordinance outlaws feeding deer.

Nothing is being done to control the population, Joslin said, and continuing to allow that unmitigated growth creates a host of issues.

Deer can threaten human safety in several ways. Bucks sometime become aggressive toward people during the fall mating season. Does at times do the same in the spring, when they are protective of their fawns.

As the numbers of deer increase, some likely will become more aggressive toward humans, she said, because the animals view people as competitors for resources, such as food and space.

They also draw predators like bears and mountain lions into city neighborhoods.

More deer also would mean more property damage, more collisions between animals and vehicles, and more health problems for the deer.

In Helena, the animals have been found with viral skin infections, ringworm and growths that blind them or prevent them from eating, Joslin said.

continued

She presented a simple population growth model. Beginning with one buck and one doe, and assuming females would produce one fawn each year, the mating pair would multiply into 120 deer in a decade. Using the same scenario, but assuming every doe gave birth to twins, the original four-legged lovers would produce a herd more than 1,000 strong in 10 years, Joslin said.

Task force member Andrew Jakes said he doesn't want the herd to outgrow its welcome—a "threshold" of tolerance exists among city residents.

Another member, Tom DeYoung, said some citizens already are intolerant of deer. He said he recently witnessed a woman throw a rock at a doe.

The task force also is ironing out questions it will use in a phone survey later this fall. About 400 people will be contacted by the University of Montana's Bureau of Business and Economic Research. The $10,000 phone survey will be paid for in part by a $7,000 grant from FWP. The task force also has $5,000 in city funds at its disposal.

Some of that money may go to Gene Hickman, a consultant and wildlife biologist, who could be enlisted to determine the size of the city's herd. In his presentation Thursday, Hickman said he counted 60 deer in the Sixth Ward during a sample survey earlier this week.

The growing population is a relatively new problem. Joslin said the deer population has been noticeably growing for about five years. Some of the dozen bucks euthanized in the city last year were 4-year-olds, and represented some of the oldest males found in the city.

When her father was growing up in Helena, she said, news of a hunter finding a deer track spread fast in the city.

Name _____   Date _____

Reasoning About a Situation or Event

What Is the Situation?

Who are the stakeholders for this situation?

What is the point of view for each stakeholder?

What are the assumptions of each group?

What are the implications of these views?

Models

5

Instructional Purpose

• To introduce the concept of models

Curriculum Alignment **Goal 1** Concept ◯ **Goal 2** Content ◯ **Goal 3** Process/ Experimental Design ◯ **Goal 4** Process/ Reasoning

---

Vocabulary

Conceptual Model A model created by making the conceptual leap of seeing similarities between something we do not understand (the thing we want to model) and something we understand well (the conceptual model)

Mathematical Model A mathematical relationship that behaves in the same way as the thing being modeled

Physical Model An actual physical device or process that behaves like the object or phenomenon being modeled but represents a simplified version

Materials/Resources

• Models (Teacher Resource 1)
• Models 1 (Handout 5.1)
• Models 2 (Handout 5.2)
• Problem Log Question (Handout 5.3)

Lesson Length

60 minutes

 Activities

1. Before you begin this lesson, read the "Models" section in *Science for all Americans* (Teacher Resource 1: 1990 edition: pages 168–172) and familiarize yourself with the generalizations this book makes about the concept.

2. Tell students that you need to interrupt their work on the problem for some direct teaching about the concept of models. Have them brainstorm

---

examples of models. Ask them to come up with a definition of the word "model"; encourage class discussion to refine their definition of the term.

3. Tell the students that models are extremely important tools for scientists as well as for other people. Tell them that they are now going to explore the concept in more detail, using examples from a topic that is interesting not only to scientists but also to other people: the human body. (This is by no means the only possible example, but it has the advantage of being very familiar to the students.) Ask them to provide examples of models that could be used to learn more about this topic. List these examples on the board. Have the students list their models on **Models 1** (Handout 5.1), or give them a copy of the class list for their **Problem Logs.**

   Examples should include some (or all) of the following: (If the computer models don't come up in the discussion, be sure to mention them.)

   • Computer simulations of the circulatory system, respiratory system, etc.
   • CPR training dummies
   • Mannequins
   • Crash test dummies
   • The overlay diagrams from an encyclopedia
   • The Visible Man/Woman (plastic three-dimensional models that students can construct from a kit)
   • Medical demonstration models
   • Medical drawings
   • Plastic hearts or other body parts
   • Photographs
   • Art; Michelangelo's *David,* for example
   • Real animals, for dissection
   • A frog dissection simulation program (accepted in lieu of real dissection experience in some high schools)
   • Artist's model: live humans
   • Cadavers
   • The Visible Human: an online database of CT, MRI and cryosection images of male and female cadavers (http://www.nlm.nih.gov/research/visible/visible_human.html)

4. Tell the students that models can be classified into three categories, as described in *Science for All Americans:* physical, conceptual, and mathematical models. Discuss the definitions of each type with them.

   Physical Model An actual physical thing (device or process) that behaves like the thing/phenomenon being modeled, but in a simplified version. (Example: A model rocket, when it is being used to study ballistic properties of real spacecraft; it looks and acts a lot like the real thing, but is smaller, costs a lot less, weighs less, cannot go as far.)

Conceptual Model A model created by making the conceptual leap of seeing similarities between something we do not understand (the thing we want to model) and something we understand well (the conceptual model). When we use analogies or metaphors to describe something, we are using conceptual models. (Example: Saying that a muscle is like a rubber band, then thinking about the properties of rubber bands and using our understanding of these to make predictions about the properties of muscles. Note that while the rubber band is itself a physical thing, it is being used here as a conceptual model of a muscle. Rubber bands could also be used directly as a physical model of a muscle.)

Mathematical Model A mathematical relationship that behaves in the same way as the thing being modeled. These can be simple (for example, distance = rate × time) or complicated (e.g. a complex computer program that is used for weather prediction).

5. Ask the students to classify the model systems they described for the human body as physical, conceptual, or mathematical models on **Models 1** (Handout 5.1). Have the students do the worksheet in small groups so that they can discuss their classifications; then, debrief the classifications and the reasons for classifying model as they did. Note that the students' justifications for their classification choices are as important as the choices they make.

Sample answers include:

• Computer simulations: mathematical, conceptual

• CPR training dummies: physical

• Crash test dummies: physical

• The overlay diagrams from an encyclopedia: physical

• The Visible Man/Woman (plastic three-dimensional models that students can construct from kit): physical

• Medical demonstration models: physical

• Medical drawings: physical

• Photographs of intact and dissected bodies: physical

• Art: Michelangelo's *David,* for example: physical

• Real animals, for dissection: while they are physical things, they could be better classified as conceptual models (because we have made the conceptual leap that animal bodies, in many important ways, are very much like human bodies).

• A frog dissection simulation program: physical, because it results in physical images; conceptual, because we are assuming that frog bodies are enough like human bodies that we can learn something useful about anatomy by studying them; and mathematical, because the model results from a computer program.

- Artist's model: physical
- Cadavers: physical; also conceptual, because we are assuming that dead people are enough like live people that we can study them to learn about how the body works.

6. Remind the students that the important thing about models is that they are useful, especially when you do not have access to the original or when you are predicting future activity. Ask the students to develop a list of people who might find a particular model useful and why, as well as why they would use the model. List the answers on the board. Answers may include:

- Artists, to create art
- Doctors working out a new surgical procedure
- Medical students learning about the human body
- Students in biology class
- Fashion designers
- Automotive safety engineers
- School bus drivers learning CPR

7. Ask the students to complete **Models 2** (Handout 5.2), matching models and their users. Once the students have done this, discuss their work. Ask them whether every model was equally useful.

8. Ask the students which of the models discussed previously would be the best for teaching medical students about the human body. Do they think that any one model would be enough to train future doctors? Why or why not? The discussion should relate to the generalization that all models have limitations, and that over reliance on a single model can be misleading because of the inherent limitations of the model. (If the students do not come up with this idea, be sure to bring it up yourself.)

9. Have a discussion regarding whether any of the models could be improved in such a way that they would be more useful.

 **Notes**

1. The background reading on the concept of models is very important. *Science for All Americans* is a primary source document that has had significant influence in setting the stage for science education reform.

2. See *Appendix A,* The Concept of Models, for more background information about using the concept of models in science.

Homework

Have students complete the **Problem Log Question** (Handout 5.3) as homework.

 # Extending Student Learning

1. Have students make a list of the models that they use in school, at home, on a job, or in a hobby or sport, and the ways in which these models are useful.
2. Have students research how models are used in medical training and in other careers.

 # Assessment

- Class discussion: Evaluate student engagement and contribution to the class discussion.
- Models 1 and Models 2: Assess student understanding of the characteristics and uses of models.
- **Problem Log Question:** Assess student understanding of the characteristics and uses of models.

 # Technology Integration

- Have students explore the Internet for examples of computer-generated models.
- Use a computer simulation program, such as a frog dissection simulation program, to provide students with additional background information for comparing the use of models to the use of actual objects/organisms.

Models
(Teacher Resource 1)

A model of something is a simplified imitation of it that we hope can help us understand it better. A model may be a device, a plan, a drawing, an equation, a computer program, or even just a mental image. Whether models are physical, mathematical, or conceptual, their value lies in suggesting how things either do work or might work. For example, once the heart has been likened to a pump to explain what it does, the inference may be made that the engineering principles used in designing pumps could be helpful in understanding heart disease. When a model does not mimic the phenomenon well, the nature of the discrepancy is a clue to how the model can be improved. Models may also mislead, however, suggesting characteristics that are not really shared with what is being modeled. Fire was long taken as a model of energy transformation in the sun, for example, but nothing in the sun turned out to be burning.

Physical Models

The most familiar meaning of the term "model" is the physical model—an actual device or process that behaves enough like the phenomenon being modeled that we can hope to learn something from it. Typically, a physical model is easier to work with than what it represents because it is smaller in size, less expensive in terms of materials, or shorter in duration.

Experiments in which variables are closely controlled can be done on a physical model in the hope that its response will be like that of the full-scale phenomenon. For example, a scale model of an airplane can be used in a wind tunnel to investigate the effects of different wing shapes. Human biological processes can be modeled by using laboratory animals or cultures in test tubes to test medical treatments for possible use on people. Social processes too can be modeled, as when a new method of instruction is tried out in a single classroom rather than in a whole school system. But the scaling need not always be toward smaller and cheaper. Microscopic phenomena such as molecular configurations may require much larger models that can be measured and manipulated by hand.

A model can be scaled in time as well as in size and materials. Something may take so inconveniently long to occur that we observe only a segment of it. For example, we may want to know what people will remember years later of what they have been taught in a school course, but we settle for testing them only a week later. Short-run models may attempt to compress long-term effects by increasing the rates at which events occur. One example is genetic experimentation on organisms such as bacteria, flies, and mice that have large numbers of generations in a relatively short time span. Another important example is giving massive doses of chemicals to laboratory animals to try to get in a short time the effect that smaller doses would produce over a long time. A

Pages 168–172 from *Science for All Americans* by R.J. Rutherford et al. Copyright © 1991. Reprinted by permission of Oxford University Press, UK.

mechanical example is the destructive testing of products, using machines to simulate in hours the wear on, say, shoes or weapons that would occur over years in normal use. On the other hand, very rapid phenomena may require slowed-down models, such as slow-motion depiction of the motion of birds, dancers, or colliding cars.

The behavior of a physical model cannot be expected ever to represent the full-scale phenomenon with complete accuracy, not even in the limited set of characteristics being studied. If a model boat is very small, the way water flows past it will be significantly different from a real ocean and boat; if only one class in a school uses a new method, the specialness of it may make it more successful than the method would be if it were commonplace; large doses of a drug may have different kinds of effects (even killing instead of curing), not just quicker effects. The inappropriateness of a model may be related to such factors as changes in scale or the presence of qualitative differences that are not taken into account in the model (for example, rats may be sensitive to drugs that people are not, and vice versa).

Conceptual Models

One way to give an unfamiliar thing meaning is to liken it to some familiar thing—that is, to use metaphor or analogy. Thus, automobiles were first called horseless carriages. Living "cells" were so called because in plants they seemed to be lined up in rows like rooms in a monastery; an electric "current" was an analogy to a flow of water; the electrons in atoms were said to be arranged around the nucleus in "shells." In each case, the metaphor or analogy is based on some attributes of similarity—but only some. Living cells do not have doors; electric currents are not wet; and electron shells do not have hard surfaces. So we can be misled, as well as assisted, by metaphor or analogy, depending on whether inappropriate aspects of likeness are inferred along with the appropriate aspects. For example, the metaphor for the repeated branching of species in the "tree of evolution" may incline one to think not just of branching but also of upward progress; the metaphor of a bush, on the other hand, suggests that the branching of evolution produces great diversity in all directions, without a preferred direction that constitutes progress. If some phenomenon is very unlike our ordinary experience, such as quantum phenomena on an atomic scale, there may be no single familiar thing to which we can liken it.

Like any model, a conceptual model may have only limited usefulness. On the one hand, it may be too simple. For example, it is useful to think of molecules of a gas as tiny elastic balls that are endlessly moving about, bounding off one another; to accommodate other phenomena, however, such a model has to be greatly modified to include moving parts within each ball. On the other hand, a model may be too complex for practical use. The accuracy of models of complex systems such as global population, weather, and food distribution is limited by the large number of interacting variables that need to be dealt with simultaneously. Or, an abstract model may fit observations very well, but have no intuitive meaning. In modeling the behavior of molecules, for instance, we have to rely on a mathematical description that may not evoke any associated mental picture. Any model may have some irrelevant features that intrude on our use of it. For example, because of their high visibility and status, athletes and entertainers may be taken as role models by children not only in the aspects in which they excel but also in irrelevant—and perhaps distinctly less than ideal—aspects.

Mathematical Models

The basic idea of mathematical modeling is to find a mathematical relationship that behaves in the same way the system of interest does. (The system in this case can be other abstractions, as well as physical or biological phenomena.) For example, the increasing speed of a falling rock can be represented by the symbolic relations $v = gt$, where g has a fixed value. The model implies that the speed of fall (v) increases in proportion to the time of fall (t). A mathematical model makes it possible to predict what phenomena may be like in situations outside of those in which they have already been observed—but only what they may be like. Often, it is fairly easy to find a mathematical model that fits a phenomenon over a small range of conditions (such as temperature or time), but it may not fit well over a wider range. Although $v = gt$ does apply accurately to objects such as rocks falling (from rest) more than a few meters, it does not fit the phenomenon well if the object is a leaf (air drag limits its speed) or if the fall is a much larger distance (the drag increases, the force of gravity changes).

Mathematical models may include a set of rules and instructions that specifies precisely a series of steps to be taken, whether the steps are arithmetic, logical, or geometric. Sometimes even very simple rules and instructions can have consequences that are extremely difficult to predict without actually carrying out the steps. High-speed computers can explore what the consequences would be of carrying out very long or complicated instructions. For example, a nuclear power station can be designed to have detectors and alarms in all parts of the control system, but predicting what would happen under various complex circumstances can be very difficult. The mathematical models for all parts of the control system can be linked together to simulate how the system would operate under various conditions of failure.

What kind of model is most appropriate varies with the situation. If the underlying principles are poorly understood, or if the mathematics of known principles is very complicated, a physical model may be preferable; such has been the case, for example, with the turbulent flow of fluids. The increasing computational speed of computers makes mathematical modeling and the resulting graphic simulation suitable for more and more kinds of problems.

Name _____   Date _____

Models 1

Model	Category	Reason for Placing in Category
1.		
2.		
3.		
4.		
5.		
6.		
7.		
8.		
9.		
10.		

Name _____   Date _____

Models 2

Model	User
A.	
B.	
C.	
D.	
E.	
F.	
G.	
II.	
I.	
J.	

continued

Which models do you think are the most useful to the user? Why do you say this?

Name _____   Date _____

Problem Log Question

Choose one of the models from the list. What improvements could be made to make it more useful? Why would these improvements be helpful?

Maps as Models

6

Instructional Purpose

• To use maps in exploring the Lyme disease problem

Curriculum Alignment Goal 1 **Concept** Goal 2 **Content** Goal 3 **Process/ Experimental Design** ◯Goal 4 **Process/ Reasoning**

 Materials/Resources

- Various maps of local area (street, topographic, tax, orthophoto, etc.)
- Ruler for each group of 3 to 4 students
- Grease pencil for each group of 3 to 4 students
- Clear plastic boxes with clear lids (such as shoe storage boxes)
- Modeling clay (oil-based; not Play Doh)
- Water pitchers and water
- Buckets for waste
- Optional: *Earthsearch: A Kid's Geography Museum in a Book,* by John Cassidy. Published by Klutz Press, Palo Alto, CA, 1994.
- *John Snow: The London Cholera Epidemic of 1854* (Handout 6.1)

- Map Questions (Handout 6.2)
- Making a Contour Map (Handout 6.3)

Lesson Length

120 minutes

 **Activities**

1. Discuss why a map is an example of a model. **Ask:**
   - How is a map similar to the area that it represents?
   - How is it different?
   - Is a map a physical, a conceptual, a mathematical model, or more than one of these?

2. Discuss how maps might be useful in resolving the problem that students are dealing with in this unit. As an example, you might use the story of how

Dr. John Snow discovered the cause of cholera by plotting known cases on a map of London. Use *John Snow: The London Cholera Epidemic of 1854* (Handout 6.1). Also see pages 24 to 25 of *Earthsearch: A Kid's Geography Museum in a Book* for additional information.

3. Have students examine various maps of your local area. Have them answer the **Map Questions** (Handout 6.2) as they study the maps.

4. As a whole group, have the students compare and contrast the maps that were used. **Ask:** Is there a standard for making maps? What kinds of maps do you think would be useful in resolving the problem in this unit?

5. Have students complete the **Making a Contour Map** activity (Handout 6.3) that demonstrates what contour maps represent.

 ## Note

Students may need some review related to maps and the use of a compass for this lesson.

 ## Homework

Have each student draw a map of his/her immediate neighborhood. Remind the students that a map should have a title, a compass rose, and a map scale.

 ## Extending Student Learning

Have students find examples of different types of maps in various books and on Internet sites. Have them discuss the effectiveness of various types of maps as models.

 ## Assessment

- Class discussion: Evaluate student engagement and contribution to the class discussion.

 ## Technology Integration

- Have students explore Google Earth (http://earth.google.com/) to view various geographic locations. Google Earth combines satellite images and maps to allow viewers to obtain geographical information about a location.

John Snow: The London Cholera Epidemic of 1854 by Scott Crosier

Background

John Snow (1813–1858) was educated at a private school until, at the age of fourteen, he was apprenticed to a surgeon living at Newcastle-on-Tyne. After serving as a colliery surgeon and unqualified assistant during the London Cholera epidemic of 1831–2, he became a student at the Huntierian School of Medicine in Great Windmill Street, London. After two years of schooling, he was accepted a member of the Royal College of Surgeons of England. He graduated M.D. of the University of London in 1844.

In 1849 Snow published a small pamphlet "On the Mode of Communication of Cholera" where he proposed that the "Cholera Poison" reproduced in the human body and was spread through the contamination of food or water. This theory was opposed to the more commonly accepted idea that Cholera, like all diseases, was transmitted through inhalation of contaminated vapors. Although he was awarded for this work, without the technology and knowledge that we have today, Snow had no way to prove his theory.

Innovation

It wasn't until 1854, when Cholera struck England once again, that Snow was able to legitimate his argument that Cholera was spread through contaminated food or water. Snow, in investigating the epidemic, began plotting the location of deaths related to Cholera (see illustration). At the time, London was supplied its water by two water companies. One of these companies pulled its water out of the Thames River upstream of the main city while the second pulled its water from the river downstream from the city. A higher concentration of Cholera was found in the region of town supplied by the water company that drew its water form the downstream location. Water from this source could have been contaminated by the city's sewage. Furthermore, he found that in one particular location near the intersection of Cambridge and Broad Street, up to 500 deaths from Cholera occurred within 10 days.

After the panic-stricken officials followed Snow's advice to remove the handle of the Broad Street Pump that supplied the water to this neighborhood, the epidemic was contained. Through mapping the locations of deaths related to Cholera, Snow was able to pinpoint one of the major sources of causation of the disease and support his argument relating to the spread of Cholera.

Snow's classic study offers one of the most convincing arguments of the value of understanding and resolving a social problem through the use of spatial analysis. Nonetheless, there is some controversy

continued

regarding whether Snow made the map prior to or after the removal of the pump handle and about the timing of this removal relative to the temporal pattern of cholera deaths.

While mapping has become a standard research approach in medical geography and epidemiology, today's researchers express the incidence of disease as a rate relative to the population or to the population within age cohorts (e.g., deaths per 1,000 population) so as to factor out the influence of population density. Using such refinements to the methods employed by Snow, mapping and spatial statistical techniques assist medical practitioners in understanding the diffusion and spread of diseases within communities and across the globe.

Name _____   Date _____

Map Questions

1. Find your house or school on each map. Was it easier to find on some maps than others? Why?

2. Can you tell from the maps which way streams and rivers flow? Explain.

3. Can you tell anything about the kind and density of vegetation in different areas by looking at any of the maps?

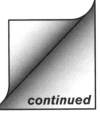

continued

4. How can you distinguish low elevations from high elevations?

5. What elements are the same on different maps? What elements are different?

6. Did any of the maps give information about where houses are located?

Name _____   Date _____

Making a Contour Map

1. Have students sculpt a "landform" out of clay in a plastic box with a clear top (such as a shoebox-size storage box). If the box does not come with a clear top, stretch plastic wrap over it.

2. Have students measure ½ inch increments on the inside of the box from bottom to top and mark these with lines with a grease pencil or marker.

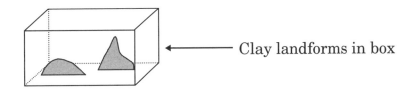 ←————— Clay landforms in box

3. Have students pour water into the box to reach the level of the first ½ inch marking. Then, have students place the clear lid on the box and looking from above, trace the outline of the water-clay interface on the lid of the box.

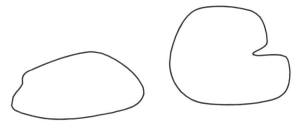

Top view of landform with ½ inch of water in box

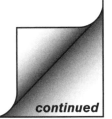

continued

4. Have students add water to reach the level of the second ½ inch marking; repeat the tracing step.

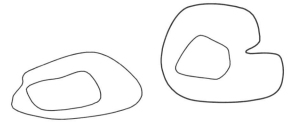

Top view of landform with 1 inch of water in box

5. Have students repeat this until the top of the landform is covered with water.

Continue to produce contour map of landform

6. When students are finished, have them use tissue or tracing paper and trace the map they have made on the box lid. Have them label each of the lines according to the depth of the water.

The Niche of Deer

7

Instructional Purpose

• To incorporate additional information about the problem in developing a solution

• To learn about the niche of deer

• To learn about the nature of human/deer interactions

• To learn the fundamental vocabulary of ecology

Curriculum Alignment ○ Goal 1
 Concept   ● Goal 2
 Content   ○ Goal 3
 Process/
 Experimental Design   ○ Goal 4
 Process/
 Reasoning

 **Vocabulary**

Buck A male deer

Community All of the organisms living in a given area

Doe A female deer

Ecology The study of ecosystems

Ecosystem A system composed of all living things (community) and the physical environment in the area

Edge The boundary between two different kinds of habitat (i.e., between forest and meadow). Edge can itself be a habitat.

Fawn A young deer, especially one still unweaned or retaining a distinctive baby coat

Habitat The kind of place where a given organism normally lives

Niche The role or position of an organism in the ecosystem, determined by its behavior and relationships to other ecosystem components. Note that a niche is not the same as a habitat.

Population Group of individuals belonging to the same species

Materials/Resources

• Resource materials about deer

• Additional Problem Statement Information A (Handout 7.1)

- Additional Problem Statement Information B (Handout 7.2)
- Problem Log Questions (Handout 7.3)

Lesson Length

120 minutes

 ## Activities: Additional Information About the Problem Statement

1. Introduce new information about the problem with **Additional Problem Statement Information A** and **Additional Problem Statement Information B** (Handouts 7.1 and 7.2). Ask students if they think the problem has changed. Does this raise any new questions? Revise the **Need to Know Board** to reflect the new information and the new questions. You should anticipate questions such as the following, but there will be many more:
   - What do deer eat?
   - How long do they live?
   - What are natural predators of deer?
   - How many offspring do deer produce each year?
   - What is the life span of deer?
   - What is the range of their habitat?
   - Do they migrate?
   - What are the social patterns of deer?
   - What are their favorite plants?
   - Are there plants that they do not eat?
   - Why are deer attracted to salt?

2. Have student groups gather information either from the classroom resource materials or from outside resources. They should report back to the class and cite sources for their information.

 ## Activities: Discussing New Information

1. Debrief student findings and update the **Need to Know Board** to reflect the new information and new questions.

2. With information about deer that has been gathered by the students, make a concept map or other model of the ecosystem of the deer. You may use questions such as the following to prompt the development of this model. **Ask:**
   - Should deer be at the center of the concept map? Why or why not?
   - What other parts of the ecosystem do deer interact with?

- What predators prey on deer?
- What parasites depend on deer?
- What are food sources for the deer?
- What physical terrain do deer require?
- Do deer prefer deep forest, forest edge, or meadow? Why?
- How many car accidents are caused each year by deer?

 ## Homework

- Have students complete the **Problem Log Questions** (Handout 7.3) for homework.
- Throughout the unit, students should continue to collect resource information that will assist them in resolving the problem.

 ## Extending Student Learning

Have students find out about another species of animal that has experienced conflict with humans with respect to its habitat. Examples include:

- Rats, pigeons, starlings
- Coyotes in Los Angeles
- Buffalo or wolves in Yellowstone
- Raccoons in Central Park

Students may share the information with their classmates in the form of a report or presentation.

 ## Assessment

- Evaluate the number, quality, and variety of sources of information students collect regarding the deer questions.
- **Problem Log Questions:** assess for student understanding of the relationship between humans and deer.

Additional Problem Statement Information A

MEMO

To: Deputy Mayor Marie Barnes

From: Jane White, Mayor

Marie,

Handle this one ASAP! Let me know how it goes.

Jane

Mayor Jane White
City Hall

Dear Ms. White:

I am writing to express my frustration and indignation with the state of affairs in this town. I recently moved here from Brooklyn, New York, and was looking forward to enjoying my retirement in a beautiful new home in the suburbs. I spent over $10,000 re-landscaping my new property. Last weekend, something ate the tops off most of my new plants. The rest of the plants were trampled. I didn't see what did this, but my next-door neighbors say that the deer are a real nuisance. They have given up on their garden.

Sincerely,

Ira B. Stanley

P.S. I've also sent a copy of this letter to *The Kingsford Gazette*.

Additional Problem Statement Information B

MEMO

From: Jane White, Mayor

To: Marie Barnes, Deputy Mayor

Subject: Here are three more!

Dear Mrs. White,

My next door neighbor has put out a salt lick that is attracting deer from everywhere! They pass through my yard on the way there and eat all of my plants! Can't you pass a law forbidding salt licks?

Your constituent,
Melissa Caldis

Dear Mayor White,

I don't know why people of this community are so intolerant of the deer. They were here first! It's not their fault that people moved in and took their habitat. I say, "live and let live."

Sincerely,
George Gunderson

Dear Mayor White,

I'm concerned with the increasing number of deer/car accidents. Susan Hollis, the Kingsford agent for State Farm Insurance, says that claims for deer-related damage have increased by more than 100 percent since 2001. Something must be done!

Martin Reardon

Problem Log Questions

1. What are some of the problems that arise when deer and humans occupy the same locality?

2. As towns and suburbs continue to grow, the number of forests and fields left for deer is reduced. How does this "growth" factor affect deer?

Introduction to Experimental Design: Deer Repellents

8

Instructional Purpose

- To introduce the principles of experimental design.
- To allow students to design an experiment to test the efficacy of deer repellents

Curriculum Alignment    Goal 1 Concept    Goal 2 Content    Goal 3 Process/ Experimental Design   ○ Goal 4 Process/ Reasoning

 Vocabulary

Constants Factors in an experiment that remain the same

Control The standard or baseline for comparing experimental effects

Dependent Variable The experimental variable that responds to changes in the independent variable; you may use more than one dependent variable

Hypothesis Educated guess about what will happen in an experiment

Independent Variable The experimental variable that you change

Scientific Process The means by which scientific research is performed and disseminated

 Materials/Resources

- Bar of bath soap
- Cheesecloth or nylon net
- Copies of school district policy on the use of animals in student research
- Letter from Uncle Lou (Handout 8.1)
- Student Brainstorming Guide (Handout 8.2)

- Experimental Design Diagram (Handout 8.3)
- Experimental Design Diagram Checklist (Handout 8.4)
- Experimental Protocol (Handout 8.5)
- Laboratory Report (Handout 8.6)

Lesson Length

60 minutes for design of experiment; experiment is ongoing

 **Activities**

1. Give the students the **Letter from Uncle Lou** (Handout 8.1). Have them read it, then ask them if they think Uncle Lou's idea could work. (Be sure they justify their answers.) Ask them how they could find out if Uncle Lou is correct.

2. Ask the students how a scientist would answer this question. If necessary, prompt them to suggest an experiment. Show them the soap and the cheesecloth. Tell them that these are the materials that your stakeholder (Chris Barnes) has available for his experiment.

3. Give the students the **Student Brainstorming Guide** (Handout 8.2). They should work in small groups to complete this; after each group has finished, you should review the guide with them and ask them to explain and justify their answers.

4. Ask each small group to describe its approach to the class. Then remind the students that the experiment they are proposing involves live deer. Ask them if it would be ethical to harm the animals during the course of the experiment. Review and use the reasoning terminology introduced in Lesson 4 as you discuss this issue. After the class discusses this, give them copies of your school district's policy for the use of animals in student research. Ask them to decide whether the experiment they would like to propose meets these requirements; if they feel it does not, ask them to make changes that will make the experiment safe for the deer.

5. Once the groups have finished the brainstorming process, give the students copies of the **Experimental Design Diagram** (Handout 8.3). Tell them that this guide will help them to put their plans into the language of formal experimental design and will also help them ensure that their experiments are well-planned. If the students are unfamiliar with the terms in this guide (or have never encountered experimental design before), explain the terms.

6. Once the students have completed their handouts, check their work with them using the **Experimental Design Diagram Checklist** (Handout 8.4). Be sure that each group has listed appropriate variables to keep constant and has included all necessary controls; be sure that each group has only one independent variable in their experiment.

7. Give each group copies of the **Experimental Protocol** (Handout 8.5). Have them complete it based on the experiment described on their **Experimental Design Diagram** (Handout 8.3). When they have finished, check their work to be sure that each experiment they have designed has appropriate controls, constants, and one independent variable; also be sure that each experiment meets the requirements for ethical treatment of animals.

8. If you have deer in your area and you have time for the students to conduct their experiments, do so. Have the students use the **Laboratory Report** (Handout 8.6) to record their results. Each student should write his or her own report using data from the group's experiment.

 Notes

1. This lesson would be a good opportunity to invite a guest speaker who is an expert on the ethical treatment of animals in research settings to speak to the students.

2. The primary purpose of this lesson is to design the experiment rather than to conduct it. If you have the time for the experiment, your students should perform it at home rather than school grounds. If there are not enough deer in the area, they might try the experiment with squirrels, blue jays, or another more frequently observed animal.

3. The following books may be used to introduce and study experimental design in depth prior to this lesson:

   Cothron, J. H., Giese, R. N., & Rezba, R. J. (2005). *Science experiments and projects for students.* Dubuque, IA: Kendall/Hunt Publishing Company.

   Cothron, J. H., Giese, R. N., & Rezba, R. J. (2000). *Students and research: Practical strategies for science classrooms and competition.* Dubuque, IA: Kendall/Hunt Publishing Company.

 Homework

Depending on time and logistics, experiments may have to be conducted outside of class. The **Laboratory Reports** should be completed outside of class.

 Extending Student Learning

1. An alternative to the experiment suggested in the lesson would be to assess the efficacy of other deer deterrents such as noise makers, fencing, lights, etc.

2. Another alternative experiment would be one that compares different brands of bar soap as deterrents to deer.

3. Point out to the students that actually doing the experiments involving deer will be difficult in terms of time and the unpredictability of deer visits to any given area. Remind them that experiments can be done on animal models as

well as with deer. Ask them to think of an animal model that could be used to test whether soap repels hungry animals. For example, hamsters, mice, or mealworms could be "asked" to choose between an apple slice placed next to a piece of soap or an apple slice far away from the bar of soap. (Hamsters and mice might require more restrictions than mealworms under the ethical treatment of animals policy because they are vertebrates.) Have the students suggest as many animal model models as they can; then ask the students to decide which model is the best in terms of how they want to use it. **Ask:**

• Which animals are most like deer?

• Which model is easiest to use?

• Is the model enough like deer that it makes sense to do the experiment?

• What kind of a model is this—conceptual, physical, or mathematical?

Once the students have developed a model system that they think will be useful, have them plan and perform their experiments as described in the plan for this lesson.

 ## Assessment

• Evaluate the whole-group discussion and small-group discussions for student understanding of experimental design.

• Review the Experimental Design Diagram, Experimental Design Diagram Checklist, Experimental Protocol, and the Laboratory Report for student understanding of experimental design.

Letter from Uncle Lou

Marie,

Here's another idea for dealing with the deer. You've got deer at your house, right? Could you give this a try and see if it works?

Jane

Dear Jane,

I enjoyed seeing you at the family reunion last weekend. I was thinking tonight about what you had said about all of the problems your townsfolk have been having with deer, and I just remembered a trick our neighbors in Kentucky used to keep the deer out of their orchard. They used to hang soap in cheesecloth bags from their trees when the apples were getting ripe—they said that the deer wouldn't go near the trees then. Thought you'd like to know!

My love to the kids,

Uncle Lou

Name _____   Date _____

Student Brainstorming Guide

1. What do we need to find out? *(What is the scientific problem?)*

2. What materials do we have available?

3. How can we use these materials to help us find out what we want to know?

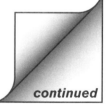

continued

Adapted from Cothron, J. G., Giese, R. N., & Rezba, R. J. (2000). *Students and research.* Dubuque, IA: Kendall/Hunt Publishing Co.

4. What do we think will happen? *(What is our hypothesis?)*

5. What will we need to observe or measure in order to answer our scientific question?

Name _____   Date _____

Experimental Design Diagram

Title of Experiment:

_____

Hypothesis (Educated guess about what will happen):

_____

Independent Variable (The variable that **you change;** there should be only one of these in your experiment):

_____

Dependent Variable (The variable that responds to changes in the independent variable; you may use more than one dependent variable):

_____

Observations/Measurements to Make:

_____

Constants (All the things or factors that remain the same):

_____

Control (The standard or baseline for comparing experimental effects):

Name _____   Date _____

Experimental Design Diagram Checklist

1. _____ Does the *title* clearly identify both the independent and dependent variables?

2. _____ Does the *hypothesis* clearly state how you think changing the independent variable will affect the dependent variable?

3. _____ Is there just one clearly defined *independent variable?*

4. _____ Is the *dependent variable* clearly stated?

5. _____ Is the expected response of the dependent variable clearly stated?

6. _____ Do the *observations / measurements* to be made include appropriate units of measure?

7. _____ Are all *constants* listed and clearly defined?

8. _____ Is there a clearly stated *control?*

9. _____ Is the experiment thoughtfully designed?

10. _____ Are safety procedures taken into consideration?

Adapted from Cothron, J. G., Giese, R. N., & Rezba, R. J. (2000). *Students and research.* Dubuque, IA: Kendall/Hunt Publishing Co.

Name _____   Date _____

Experimental Protocol

1. List the materials you will need.

2. Write a step-by-step description of what you will do (like a recipe!). List every action you will take during the experiment so that it can be repeated.

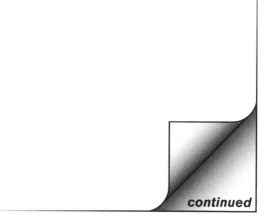

continued

3. What data will you be collecting?

4. Design a data table to collect and analyze your information.

Name _____   Date _____

Laboratory Report

1. What did you do or test? (Include your experiment title.)

2. How did you do it? What materials and methods did you use?

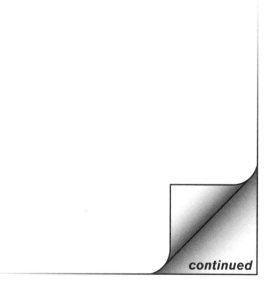

continued

3. What did you find out? (Include a data summary and the explanation of its meaning.)

4. What did you learn from your experiment?

continued

5. What additional questions do you now have?

6. How does the information you learned help with the problem?

Designing a Data Table

9

Instructional Purpose

• To understand how to construct a data table

Curriculum Alignment
 ◯ Goal 1 Concept
 ◯ Goal 2 Content
 ⬤ Goal 3 Process/ Experimental Design
 ◯ Goal 4 Process/ Reasoning

 ## Materials/Resources

• Graph Paper (Handout 9.1)
• Problem Logs

Lesson Length

60 minutes

 ## Activities

1. Have students open their **Problem Logs** to the handouts from Lesson 8. Explain to students that they are going to create a data table to organize the data they collected. Ask students to generate a title for their data table.

2. Tell students to divide a piece of paper in their **Problem Logs** in half with two columns: one for the substance used and one for the deer reaction. For any of the experiments conducted, have students identify the independent variable and the dependent variable. Tell students to complete the data table using the information from the **Experimental Protocol** (Handout 8.5) and **Laboratory Report** (Handout 8.6) in the previous lesson.

3. Distribute the **Graph Paper** (Handout 9.1). Have students write the graph title at the top of the page. Explain to students that the x axis represents the independent variable and the y axis represents the dependent

variable. Have students title both the x and y axes. Have students graph the results of their data table on the Graph Paper. Discuss as a class why a scientist records data in a table and graphs results. (Note: This activity may not be appropriate for all experiments.)

Notes

1. More information about data table construction may be found in the *Implementation Guidelines*, section 4.
2. There are many ways to present the data; the table given here is one option.

Sample Data Table

Title: Substances that repel deer

Independent Variable	Dependent Variable				Comments
	Trials				
soap					
vinegar					
rotten apple slices					
sour milk					
raw fish					

Homework

If students completed additional experiments from Lesson 8 at home, have them design corresponding data tables.

Assessment

- Data table: Evaluate for correct terminology usage and correct completion of table.
- Graph: Assess the correctness of labeling and plotting of points on the graph.

Technology Integration

- The data tables and corresponding graphs may be constructed using spreadsheet software.

Name _____ Date _____

Graph Paper

Graph Title: _____

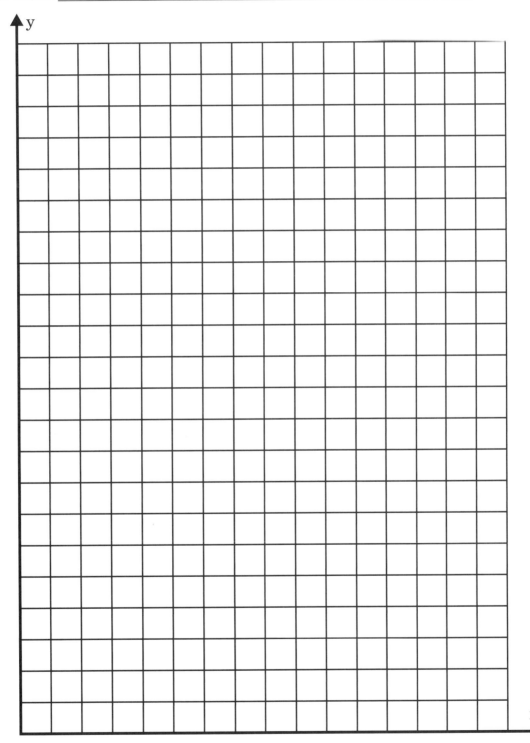

Sampling/ Estimating

10

Instructional Purpose

• To learn about different methods of estimating the size of animal populations through various sampling techniques

---

Curriculum Alignment
 Goal 1
Concept
 Goal 2
Content
 Goal 3
Process/ Experimental Design
 Goal 4
Process/ Reasoning

One of the questions that should have been asked by students on the Need to Know Board is "How many deer live in this area?" This lesson attempts to explore that question by learning how ecologists estimate animal populations. Since students will not be able to actually try tag and recapture with the deer in their area, they will simulate the process with a model (Styrofoam peanuts or lima beans).

 ## Materials/Resources

• Box of 500–1,000 Styrofoam peanuts or lima beans
• Optional: Videotape of shark estimation from *The Challenge of the Unknown*
• How Many Turtles in the Pond? (Handout 10.1)
• Key: How Many Turtles in the Pond? (Teacher Resource 1)
• Tag and Recapture I (Handout 10.2)
• Key: Tag and Recapture I (Teacher Resource 2)

• Tag and Recapture II (Handout 10.3)
• Key: Tag and Recapture II (Teacher Resource 3)
• Problem Log Questions (Handout 10.4)

(**Note:** Using *The Challenge of the Unknown Teacher's Guide* is ideal but the videotapes are out of print. Check with your school district media services department to see if these materials are owned by the school system. If not, check online secondhand booksellers, state libraries, or local PBS stations. The materials are edited by H. C. Maddux and published in 1986 by W. W. Norton and Company. If you can find only teacher's guide, that may be sufficient. Some teachers have used *Jaws of the Pacific,* a Discovery Channel DVD with similar information that is available through online secondhand booksellers.)

Lesson Length

60 minutes

 **Activities**

1. Return to the **Need to Know Board** and highlight any questions dealing with how many deer actually live in the area. **Ask:**

   • What questions do we have on the "Need to Know" board that relate to the size of the deer population?

   • How can we find out answers to these questions?

   • If we call local wildlife experts and ask them how they estimate animal populations, what do you think they will tell us?

   Brainstorm a list of possible ways that you think the deer population in a specified area can be estimated. Tell students about the homework assignment, emphasizing that they will be finding out from experts how actual counts are taken in the field.

2. In order to introduce some estimation activities, have students estimate the number of styrofoam peanuts in a box. Ask them to write a paragraph describing how they determined their estimate.

3. If possible, show the video clip about the estimation of sharks (about five minutes) from the estimation portion of *The Challenge of the Unknown.* In the film Sam Gruber, a biologist who studies sharks, is tagging and recapturing lemon sharks in order to estimate the size of the population in the Bimini Lagoon. **Ask:**

   • Why do you think Sam Gruber is interested in the number of sharks in the Bimini Lagoon?

   • Do you think he gets paid to do this? How do you know? Who do you think pays him? (**Note to teacher:** This will provide an opportunity to talk about how scientists get funded, if you wish to engage in this discussion.)

   • What models is Sam Gruber using in his work?

   • Why is Sam Gruber unsure of his estimate from the tag and recapture?

   • What is the basic math concept behind the tag and recapture method? (proportional reasoning)

4. Engage students in further investigation of "tag and recapture" as an estimation technique.

   a. Distribute **How Many Turtles in the Pond?** (Handout 10.1). Ask students to work the problem individually and then share their ideas in small groups.

   b. Discuss the tag and recapture technique with the whole class, and distribute **Tag and Recapture I** (Handout 10.2) to the small groups.

   c. In groups, have students design a way to test the accuracy of tag and recapture. Since they do not have an actual animal population to try the test on, they will have to use a model. (Students may use the box of styrofoam that was estimated at the beginning of the lesson, or lima beans, or any other readily available materials.) Distribute **Tag and Recapture II** (Handout 10.3). It would be a good idea to approve the test designed in Step 1 of the handout before conducting it.

   d. Debrief the designs students recorded in Handout 8.3 and discuss the conclusions for various groups.

   e. Have students complete the **Problem Log Questions** (Handout 10.4). This set of questions is intended to make students consider the differences in strategies in estimating a domestic animal population and a wild animal population.

Note

Here is a sample protocol for the test in Handout 10.3:

1. Obtain a box of Styrofoam packing peanuts. Remove 50 peanuts and mark them with a colored marker. Return them to the box and mix them up.

2. "Recapture" 80 peanuts by removing them one at a time (without looking). Count the number that have been marked.

3. Set up and solve the proportion: $\dfrac{\text{\#tagged on recapture}}{\text{\# recaptured}} = \dfrac{\text{\# tagged originally}}{\text{\# of total population}}$

$$\frac{\text{\# tagged on recapture}}{80} = \frac{50}{x}$$

4. Redo the recapture two more times, and see if the results are similar to the first time.

5. Count all the peanuts in the box. Compare this number to the results from number 3 above.

Homework

Have students contact any of the local or state agencies that might have information about the size of local or state-wide deer populations. Have them try to obtain statistics about the size of these groups now, over the last few years, and over the long term. Have them research the method used to gather these data. Students may prepare written or oral summaries of their findings.

 ## Extending Student Learning

Have students complete these activities to extend their learning about estimating population sizes:

- Design a way to estimate the population of rabbits on a farm. What are the advantages and disadvantages of your method?
- Think of a way to count a real animal population (e.g., ants in one anthill, snails in your garden, hummingbirds, Japanese beetles, or crickets in your back yard) and do it.

 ## Assessment

- How Many Turtles in the Pond?: may be used as a pre-assessment.
- Tag and Recapture activities: evaluate for student understanding of the estimating technique.
- **Problem Log Questions:** assess for student understanding of estimating techniques.

 ## Technology Integration

- Students may record all data in electronic format.

Name   Date

How Many Turtles in the Pond?

An ecology graduate student is estimating the number of turtles that live in Lake Matoaka. She catches 42 turtles on the banks of the lake, marks their shells, and releases them. The next week she returns to the lake where she catches 30 turtles on the banks of the lake. Of the 30 caught, 5 are marked.

What can you conclude about the number of turtles living around the pond? Explain. Show your work.

Key: How Many Turtles in the Pond?
(Teacher Resource 1)

An ecology graduate student is estimating the number of turtles that live in Lake Matoaka. She catches 42 turtles on the banks of the lake, marks their shells, and releases them. The next week she returns to the lake where she catches 30 turtles on the banks of the lake. Of the 30 caught, 5 are marked.

    What can you conclude about the number of turtles living around the pond? Explain. Show your work.

There should be about 252 turtles.

$$\frac{\text{part (recapture)}}{\text{total (recapture)}} = \frac{\text{part (whole population)}}{\text{total (whole population)}}$$

$$\frac{5}{30} = \frac{42}{x}$$
$$5x = 1260$$
$$x = 252$$

The proportion of tagged turtles in the recaptured sample should approximate the proportion in the whole population.

Name _____ Date _____

Tag and Recapture I

$$\frac{\text{\# tagged on recapture}}{\text{\# caught on recapture}} = \frac{\text{total number tagged}}{\text{total number in population}}$$

1. You tagged 90 sharks on Monday. You then caught 100 on Tuesday and found that 25 of them were tagged. What would you estimate the total population to be? Show your work.

2. Is there a different proportion you could set up to solve this problem?

3. How could you improve the accuracy of your estimate?

4. What generalizations can you make about when tag and recapture is an appropriate method to use? (Can you count any animal population this way?)

Key: Tag and Recapture I
(Teacher Resource 2)

$$\frac{\text{\# tagged on recapture}}{\text{\# caught on recapture}} = \frac{\text{total number tagged}}{\text{total number in population}}$$

1. Suppose you tagged 90 sharks on Monday. You then caught 100 on Tuesday and found that 25 of them were tagged. What would you estimate the total population to be? Show your work.

$$\frac{25}{100} = \frac{90}{x} \qquad 25x = 9000 \qquad \text{(Or, note that the ratio is 1:4.)}$$
$$x = 360$$

2. Is there a different proportion you could set up to solve this problem?

$$\frac{25}{90} = \frac{100}{x} \qquad \text{or} \qquad \frac{90}{25} = \frac{x}{100} \qquad \text{or} \qquad \frac{100}{25} = \frac{x}{90}$$

3. How could you improve the accuracy of your estimate?

   • Tag more sharks
   • Recapture more sharks

4. What generalizations can you make about when tag and recapture is an appropriate method to use? (Can you count all animal populations this way?)

   • When you cannot count the animals directly.
   • Population must be static (e.g., not in the process of migration). A well-defined habitat (such as a lake) is ideal.
   • You must be able to catch the animals and tag them.

Name _____   Date _____

Tag and Recapture II

Design a way to test the accuracy of tag and recapture. You will have to use a model since we do not have an animal population readily available for this test.

1. Describe your model and your test. How is your model similar to a real animal population? How is it different?

2. What are the advantages and disadvantages of the model you are using for the test?

3. Conduct the test. What are your conclusions about the accuracy of tag and recapture as an estimation technique?

4. Do you think that your findings can be applied to animal populations in the wild?

Key: Tag and Recapture II
(Teacher Resource 3)

Design a way to test the accuracy of tag and recapture. You will have to use a model since we do not have an animal population readily available for this test.

1. Describe your model and your test. How is your model similar to a real animal population? How is it different?

   (Answers may vary.) Use a set of items such as beans as the model. Do the tag and recapture to estimate the total bean population. Then count the actual number of beans. Repeat the tag and recapture several times to see how similar the results are. See how close the average comes to the actual number of beans.

2. What are the advantages and disadvantages of the model you are using for the test (as compared to using real animals)?

   Advantages:
   The model is easier to tag and count than an animal population.
   No animal ethics issues.

   Disadvantages:
   Because beans do not move around, they do not truly model what it is like to catch an animal.

3. Conduct the test. What are your conclusions about the accuracy of tag and recapture as an estimation technique?

   (Answers will vary depending on student results. Usually it is a pretty good estimator if a large enough sample is tagged and a large enough sample is recaptured.)

4. Do you think that your findings can be applied to animal populations in the wild?

   The model probably gives estimates that are more accurate than when used in the wild due to the fact that the animals are not uniformly distributed and they move around.

Handout 10.4

Name _____   Date _____

Problem Log Questions

1. Describe a way to estimate the population of dogs kept as pets in your school attendance area.

2. What are the advantages and disadvantages to this method? How accurate do you think this method will be?

3. How is this estimation technique similar or different from tag and recapture? How is it similar or different from the transect method that Sam Gruber used in the video?

Mathematical Model of a Deer Population

11

Instructional Purpose

• To write a mathematical model of the size of a deer population

Curriculum Alignment  Goal 1 *Concept*  Goal 2 *Content* Goal 3 *Process/ Experimental Design* Goal 4 *Process/ Reasoning*

Vocabulary

Biotic Potential The inherent capacity of an organism to increase in number; the maximum potential population growth rate

Carrying Capacity The maximum population that the environment is capable of sustaining in the long run

Environmental Resistance The sum total of the environmental limiting factors acting on a population (predation, lack of food, etc.)

Materials/Resources

• Graphing calculators
• Graph paper

• Population Model Activity (Handout 11.1)
• Problem Log Questions (Handout 11.2)

Lesson Length

60 minutes

 Activities

1. In order to construct a model of the size of the deer population in a particular area, gather the following information if it has not already been determined by prior research:

   • How many fawns does a doe have each year?

   • When does a doe reach reproductive age?

- For how many years does a doe reproduce?
- What is the average life span of a deer?

2. Identify assumptions about other aspects of deer populations (e.g., all fawns survive, that all does mate, that half the fawns are male and half are female, that all deer live to the average age and then die.) Ask why it is important to identify assumptions.

3. Have students complete the **Population Model Activity** (Handout 11.1) based on the information they have found and their assumptions. Have them graph the data on graph paper.

4. Discuss the activity with the entire class after they have had enough time to complete it.

5. Have students complete another table that assumes a shorter life span due to hunting. Have them graph the new data on the same graph paper as the first set of data. Compare the results.

Homework

Have students complete the **Problem Log Questions** (Handout 11.2) for homework.

Extending Student Learning

Have students complete the following activities:

- Consider the following rabbit population. Create a model to help you solve the problem.

  Suppose that you start with a pair of rabbits (one male and one female) that are born on January 1. Assume that all months have equal length and that rabbits begin to produce their own young two months after their own birth. After reaching the age of two months, each pair produces a mixed pair (one male and one female) and then another pair each month thereafter. Assume no rabbits die. How many pairs of rabbits will there be after one year? (The answer is 144 pairs of rabbits.)

• Analyze the following conceptual model that suggests a relationship of stress to population density. Do you think this model applies to deer?

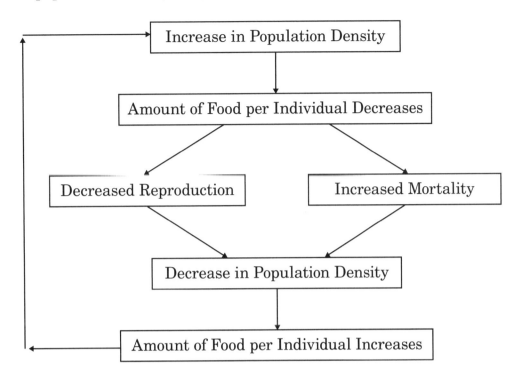

Assessment

• Population Model activity: Evaluate student understanding of how to develop a mathematical model of a population.

• Problem Log Questions: Assess student understanding of mathematical modeling.

Name _____ Date _____

Population Model Activity

1. Using the assumptions that you determined in class and the information you found on deer reproduction, complete this table, assuming the population in year one is 100.

Year	Deer Population
1	100
2	
3	
4	
5	
6	
7	
8	
9	
10	
11	
12	

continued

2. Do you think that the population would continue to increase indefinitely? What factors might limit growth?

3. Graph your totals on graph paper. (Be sure to label years on the horizontal axis and number of deer on the vertical axis.)

4. What patterns do you see in the table and your graph? Use the pattern to predict the total number of deer after 15 years. After 20 years?

Name _____   Date _____

Problem Log Questions

1. Identify the factors that affect the population model in the real world.

2. Discuss how mortality associated with organism/human interactions (such as car accidents) affect the model.

3. How might the model change if the organism is smaller in size than deer or occupies a habitat undesirable to humans?

Exponential Growth Model

Instructional Purpose

- To investigate long-term growth rates in various populations and write mathematical models to represent them

---

Curriculum Alignment  Goal 1
Concept  Goal 2
Content  Goal 3
Process/
Experimental Design ◯ Goal 4
Process/
Reasoning

Vocabulary

Exponential Growth A rate of growth characterized by rapid increases that get larger over time. Typically exponential growth functions involve a mathematical model of the form $y = (constant)^x$, such as $y = 2^x$ (the variable is used as the exponent of a constant base).

Materials/Resources

- Data regarding deer populations over a number of years in your locality or state
- Graphing calculators
- Graph paper
- Hyacinths in the Pond (Handout 12.1)
- Key: Hyacinths in the Pond (Teacher Resource 1)

- Graphing Exponential Models (Handout 12.2)
- Graph of World Population (Teacher Resource 2)
- Problem Log Questions (Handout 12.3)

Lesson Length

60 minutes

 Activities

1. Engage students in discussion about the activity in Lesson 11. **Ask:**
   - Do you think this reflects what happens in the real world? Why or why not? What can we do to find out how real populations change over time?

2. Since it is hard to collect some animal data directly, the

127

students can look at data gathered by other people or can use another organism as a model to see what we can find out about population growth. Some possible examples include:

a. Contact the game department to obtain data on deer populations in your locality or state over a number of years. Have students graph the data.

b. As an exercise in mathematical modeling, have students try the **Hyacinths in the Pond** (Handout 12.1).

c. Have students talk to someone who has raised mice or guppies. They should ask questions related to the patterns of population growth they noticed.

d. Have students use a simple experimental design to model population growth:

   • They should design a way to measure the growth of yeast or bacteria and collect the data. (Have them use the experimental design handouts from Lesson 8.)

   • Have students graph the data. **Ask:**

      • How is your graph similar to the water hyacinth graph? Can you write a mathematical model that represents the growth of the yeast or mold population?

3. After doing at least one of these activities, ask students: What are the patterns you noticed in each of the examples above? Explain the term "exponential growth" that is used to indicate this kind of pattern. Explain that a mathematical model for the hyacinth problem would be $y = 2^x$. Have students complete **Graphing Exponential Models** (Handout 12.2).

4. Have students find data on the global human population. **Ask:**

   • How has it changed over the last 50 years? The last 100? 200? How is the population changing in various countries, for example, the United States, Japan, or India? Why do these trends look different? What do you think will happen to the number of people in each country over time?

5. Show students the **Graph of World Population** (Teacher Resource 2) on planet Earth over human history. **Ask:**

   • How does the pattern compare to the patterns studied in this lesson? How can you account for the tiny valleys in the graphs?

Virginia Deer Harvest
1924–2004

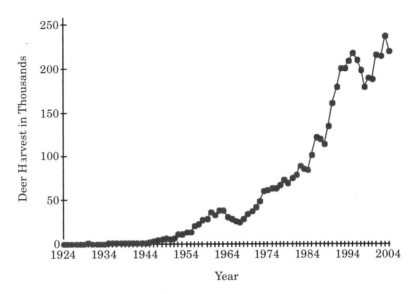

Source: Virgina Department of Game and Inland Fisheries.

Note

If you do not have actual deer data available for use with Handout 12.3 you may use the graph of deer harvest. The data for the graph were obtained from the Virginia Department of Game and Inland Fisheries. The numbers from 1947 to 2004 are obtained from hunting check station totals. Prior to 1947, the numbers are estimates.

Homework

Have students complete the **Problem Log Questions** (Handout 12.3) for homework.

Extending Student Learning

Provide these problems for students:

1. Suppose a mosquito colony had 1,500 mosquitoes in it on Monday. After 24 hours it had 2,400 mosquitoes. At this rate, estimate what the population would be after 3 days. Explain your reasoning. *(The answer, as calculated by an exponential growth formula that is beyond the scope of this grade level, is 6,944; accept answers between 6,000 and 7,500 if correct reasoning is present. Common error: this should not be done as a linear proportion. Remembering that population growth is exponential, students might reason from an*

approximation of a graph or by seeing that it increased by a factor of 1.6 in one day.)

2. The population of a suburb doubled in an 18-month period. If this growth continues and the population is now 8,000, what will the population be in 4 years? *(The answer, as calculated by an exponential growth formula that is beyond the scope of this grade level, is 50,797. If students remember that population growth is exponential rather than linear and sketch a graph from the given information, they should be able to interpolate a good approximation.)*

3. The Fibonacci numbers are the series (0, 1, 1, 2, 3, 5, 8, 13, . . .). After the two initial values, each number is the sum of the two preceding numbers. If you are interested in introducing or reviewing the concept of Fibonacci numbers in connection to population growth, you can use this problem:

Consider the following rabbit population. Create a model to help you solve the problem.

Suppose that you start with a pair of rabbits (one male and one female) born on January 1. Assume that all months have equal length and that rabbits begin to produce their own young two months after their own birth. After reaching the age of two months, each pair produces a mixed pair (one male and one female) and then another pair each month thereafter. Assume no rabbits die. How many pairs of rabbits will there be after one year? *(The answer is 144. This is an old and famous problem. The numbers of rabbits in each month form a sequence of numbers named after an Italian mathematician who went by his nickname Fibonacci. You can find information about this problem on the Internet by using the search terms "Fibonacci" and "rabbits.")*

 Assessment

- Hyacinth in the Pond problem and Graphing Exponential Models sheet: Evaluate student understanding of exponential growth models.

- **Problem Log Questions:** Assess student ability to use data from exponential growth models to make predictions and draw conclusions.

Hyacinths in the Pond

Consider the following problem:

There are water hyacinths growing in a pond on your farm. On March 1 there is one hyacinth in the pond. The number of water hyacinths in the pond doubles every week and none die.

1. Make a table showing the number of water hyacinths in the pond from March 1 to June 28.

2. Sketch a graph of your data on graph paper. (Remember that the input or independent variable corresponds to the horizontal axis.)

3. Can the pond sustain this growth forever? Explain your reasoning.

Key: Hyacinths in the Pond
(Teacher Resource 1)

Consider the following problem:

There are water hyacinths growing in a pond on your farm. On March 1 there is one hyacinth in the pond. The number of water hyacinths in the pond doubles every week and none die.

1. Make a table showing the number of water hyacinths in the pond from March 1 to June 28.

Mar. 1	1	April 5	32	May 3	512	June 7	16,384
8	2	12	64	10	1024	14	32,768
15	4	19	128	17	2048	21	65,536
22	8	26	256	24	4096	28	131,072
29	16			31	8192		

2. Sketch a graph of your data on graph paper. (Remember that the input or independent variable corresponds to the horizontal axis).

   (Should be an exponential curve)

3. Can the pond sustain this growth forever? Explain your reasoning.

   No, the carrying capacity of the pond will be reached and resources will become too scarce for the hyacinths to reproduce and sustain life.

Name _____   Date _____

Graphing Exponential Models

1. The hyacinth problem uses an exponential growth model that looks like this: $y = 2^x$.
   Graph this function on a graphing calculator with the window settings as follows:

   > xmin = –1
   > xmax = 20
   > ymin = –1
   > ymax = 20

2. On the same set of axes, graph the following function: $y = (1.5)^x$
   How does it compare to the first function?

3. Predict the shape and placement of the following functions. Then graph them on the
   same set of axes as the others.

   a. $y = (1.1)^x$
      Prediction:

      Were you correct?

   b. $y = 3^x$
      Prediction:

      Were you correct?

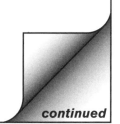

continued

4. Clear the graphs except for the graph of y = 2ˣ. Predict the placement and shape of this function and then graph it on the calculator.

    y = $(0.5)^x$
    Prediction:

    Were you correct?

Place the cursor near the y-axis and ZOOM in to get a closer look. How do the two graphs compare? What do you think explains the difference?

Graph of World Population
(Teacher Resource 2)

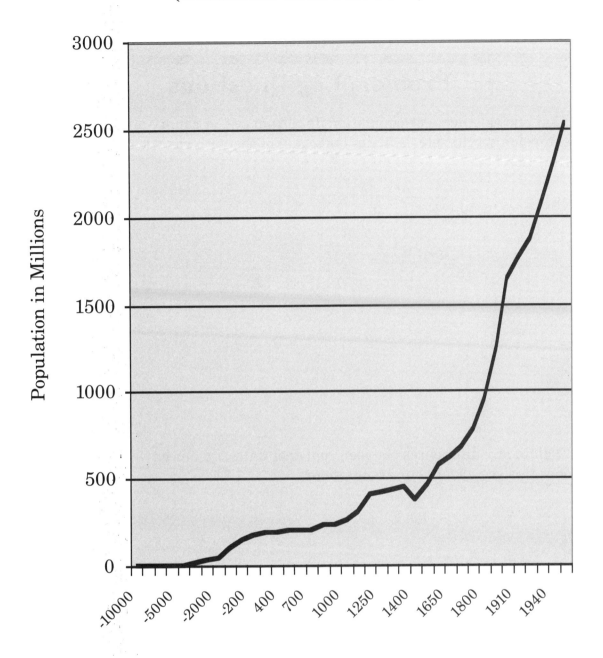

Name _____   Date _____

Problem Log Questions

1. Examine existing deer population data from city or county agencies, the extension service, etc. Is it similar to any of the patterns we have examined? If so, how?

2. Explain any differences between real deer data and our growth models. What does this say about your assumptions?

continued

3. What are the factors that could keep any population from continuing to grow exponentially?

4. What are some ways that growth is kept under control

    a. for deer?

    b. for communities?

    c. for mosquitoes?

The Spread of a Rumor: Logistic Growth Model

13

Instructional Purpose

• To use a mathematical model to model logistic growth

• To introduce the concept of exponential growth and carrying capacity

---

Curriculum Alignment

 Goal 1
Concept

 Goal 2
Content

 Goal 3
Process/
Experimental Design

 Goal 4
Process/
Reasoning

 ## Vocabulary

Logistic Growth Model A model of population growth that acknowledges that the rate of population increase may be limited by environmental factors.

Point of Inflection of a Graph The point where the curve changes from concave down to concave up or vice versa. For example, the dot on each curve below is the point of inflection.

 ## Materials/Resources

• Graphing calculators
• Rumor Record-Keeping Cards (Teacher Resource 1)
• Rules for the Rumor Activity (Teacher Resource 2)
• Data Table and Graph of the Logistic Equation (Handout 13.1)
• Graph of the Logistic Equation (Teacher Resource 3)
• Mathematical Model for Logistic Growth (Teacher Resource 4)
• Guppy Activity (Handout 13.2)

Lesson Length

120 minutes

 **Activities**

1. Rumor Activity

   a. Prepare the **Rumor Record-Keeping Cards** (Teacher Resource 1) and cut them apart so that one card can be given to each student. Mark one of the cards with the letter R in the space to the right of the zero. This symbol means that the person holding this card knows the rumor.

   b. Read this introduction to students:

      In this activity we will model the spread of a rumor by using cards that keep track of who knows the rumor. Each person will receive a card. One person "knows the rumor" when the activity begins. If there is an R next to the zero, you know the rumor. An N means you do not know the rumor.

      Explain the rules as given in **Rules for the Rumor Simulation** (Teacher Resource 2). It is wise to show a transparency as well as verbally explain the rules. Ask if there are any questions.

      Have students use their own words to explain the rules to each other in their groups to make sure they all have the same understanding. Then ask one student to explain to the whole class. (If students do not follow the directions, the activity is flawed and the data are not valid. Therefore it is imperative that you make sure they all have the correct understanding of the rules before you begin.)

   c. Distribute the cards (face down so no one observes who has the rumor). Tell students:

      When I call "Round 1," you are to stand slowly. You will have about 30 seconds to mingle and show your card to one *other person.*

   d. Call "Round 1" and when you think all students have shown their cards, tell them to return to their seats and record an R or an N depending on whether they know the rumor.

   e. Repeat the procedure for Round 2, Round 3, etc. You want to continue for at least as many rounds as it takes so that everyone has seen a rumor card. For a classroom of about 32 students, this might take about 8 or more rounds. You can stop the group after 9 rounds and quickly circulate around the room to take a peek at their cards. When you are satisfied that everyone has heard the rumor, stop calling new rounds. It does not hurt to call more rounds than necessary.

2. Collect the data in a table and create a corresponding graph.

   Ask the person who knew the rumor in Round 0 to stand. Have that person sit. Ask the people who learned the rumor on Round 1 to stand; count them, record the number in the table and have them sit. Repeat this for all rounds.

Each group should record the data in the table on **Data Table and Graph of the Logistic Equation** (Handout 13.1) and create the corresponding graph. Before students plot the data, discuss the way they labeled the axes and units.

3. Analyze the data.

   Discuss the graphs in a whole group discussion. **Ask:**

   a. Should the points on the graph be connected? In other words, are the data "continuous"? *(Not really since the data represent discrete events, but you can lightly trace the curve that is formed to see the shape. Animal population graphs would be continuous as the increases are occurring at all moments in time.)*

   b. What is the maximum value of any point on the graph? Why? *(The number of people in the simulation)*

   c. How is this graph similar to graphs we made earlier in this unit? *(the first part has the shape of an exponential growth curve)* How is it different? *(There is a point at which the graph turns from concave up to concave down; it "levels off" at the end rather than continue to increase; this is called the inflection point.)*

   d. By studying the graph, tell when you think the rumor is spreading the fastest. *(At the inflection point)*

   e. Why do you think this graph is different from our earlier graphs? *(The number of people in the simulation group is finite)*

   f. Describe in words how the growth of the rumor changes over time. *(It spreads slowly at first, speeds up to a maximum rate at the round of the inflection point and then slows down until it levels off when everyone knows the rumor.)*

   g. Tell students that the graph of the spread of a rumor is similar to the graph of population growth over a long period of time. **Ask:**

      • Can the population of deer continue to grow exponentially forever? *(No, they will begin to compete for resources)*

4. Tell students that ecologists have a mathematical model for this pattern of growth. They call it a logistic growth model. The graph of this mathematical equation starts out in the pattern of an exponential growth curve and then slows down until it levels off at the carrying capacity. Show **Graph of the Logistic Equation** (Teacher Resource 3) for a graph. Point out the features that are labeled. **Ask:**

   • How does this graph compare to the graph of the data that we collected on the spread of a rumor?

5. Show a transparency of the **Mathematical Model for Logistic Growth** (Teacher Resource 4) of the mathematical model that is commonly used to model this kind of growth. **Ask:**

- How do you think this model relates to the deer population problem?
- What factors might affect the carrying capacity for a population of deer?
- Do you think the carrying capacity now might be different from 100 years ago? Explain why you think that.
- If you look at the deer population graph in Virginia, does it look like a logistic curve? What do you think explains why it is not?
- How might population growth models for a state or country be different for a town or small community?

 Note

Ideally you should have 32 or more people participate in the rumor modeling activity. If your class is small, see if you can get another group to join your class for the rumor-spreading part of the activity.

 Homework

Have students complete **Guppy Activity** (Handout 13.2) for homework.

 Extending Student Learning

Have students complete this problem:

In the rumor data collection activity, graph the data from the second column of the data table "# of people who learned the rumor on this round."

- How does the increasing part of this graph relate to your graph in Handout 10.2?
- How does the decreasing part of this graph relate to your graph in Handout 10.2?
- On what round does the maximum increase occur? How does this relate to the inflection point on your original graph?

 Assessment

- Guppy Activity: Evaluate student understanding of logistic growth.

Rumor Record-Keeping Cards
(Teacher Resource 1)

Copy this page and cut into six cards. You will need one card for each person. You need to write an R next to the zero on one copy of the first card. Write "N" on any others that do not have a letter given for Round 0.

Round number	Do you know the rumor?
0	
1	
2	
3	
4	
5	
6	
7	
8	
9	
10	
11	
12	

Round number	Do you know the rumor?
0	N
1	
2	
3	
4	
5	
6	
7	
8	
9	
10	
11	
12	

Round number	Do you know the rumor?
0	N
1	
2	
3	
4	
5	
6	
7	
8	
9	
10	
11	
12	

Round number	Do you know the rumor?
0	N
1	
2	
3	
4	
5	
6	
7	
8	
9	
10	
11	
12	

Round number	Do you know the rumor?
0	N
1	
2	
3	
4	
5	
6	
7	
8	
9	
10	
11	
12	

Round number	Do you know the rumor?
0	N
1	
2	
3	
4	
5	
6	
7	
8	
9	
10	
11	
12	

Rules for the Rumor Simulation
(Teacher Resource 2)

1. The game is played in silence except for round numbers given by the teacher. A person "speaks" by showing his/her card. A card with R on it means that the person knows the rumor. A card looks like this.

Round number	Do you know the rumor?
0	
1	
2	
3	
4	
5	
6	
7	
8	
9	
10	
11	
12	

2. Each person MUST "speak" to EXACTLY one person in each round of the game.

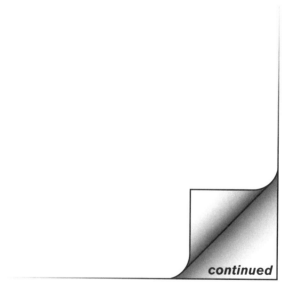

continued

3. If a person "speaks" to someone, that person may or may not choose to "talk" back.

4. A person may have one, many, or no people "speak" to her in any round.

5. At the END of a round, each person returns to his seat and records whether he found out the rumor. Once the rumor has been recorded, that person will always know the rumor.

Name _____   Date _____

Data Table and Graph of the Logistic Equation

1. Record the data in this table.

Round #	# of people who learned the rumor on this round	Total # of people who know the rumor
0	1	1
1		
2		
3		
4		
5		
6		
7		
8		
9		

2. Graph your data on the axes below.

   a. Label the horizontal axis with an appropriate title and mark the units on the axis.

   b. Label the vertical axis with an appropriate title and mark the unit on the axis.

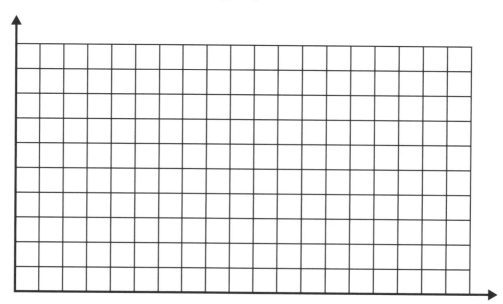

Graph of the Logistic Equation
(Teacher Resource 3)

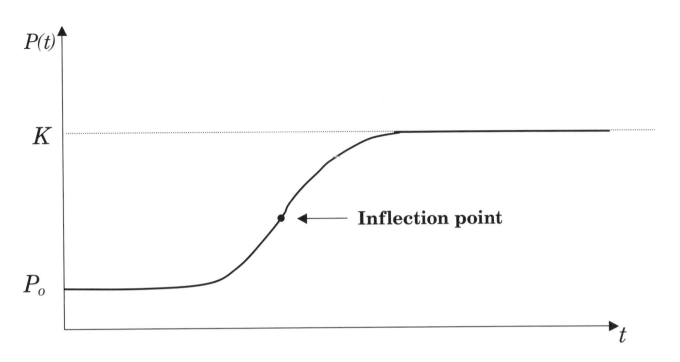

$P(t)$ is the population at time t

K is the carrying capacity

P_0 is the initial population

(Note that the inflection point typically occurs when the population is half of the carrying capacity.)

Mathematical Model for Logistic Growth
(Teacher Resource 4)

The mathematical model for logistic growth can be stated as follows

$$P(t) = \frac{K}{1 + Ae^{-kt}} \quad \text{where} \quad A = \frac{K - P_0}{P_0}$$

where

$P(t)$ is the population at time t

K is the carrying capacity

k is a constant called the relative growth rate

P_0 is the initial population

e is a constant that is an irrational number and therefore has an infinite number of digits in its decimal expansion; it is approximately 2.72

Note that P and t are the only variables in the equation. P is the dependent variable and t is the independent variable. In other words, P (the output) depends on t (the input).

Name _____ Date _____

Guppy Activity

Suppose you buy a very large aquarium and put 10 guppies in it. Two weeks later you find that there are 25 guppies in the aquarium. Then in another two weeks you find that there are 56 guppies. You start to keep records as shown in the following chart. You are worried about how many guppies can live in the aquarium.

Time t in weeks	Population $P(t)$
0	10
2	25
4	56
6	103
8	148
10	177
12	
14	
16	
18	

1. Assume that the population can be modeled by the following function. Verify that when $t = 0$, the formula gives 10 guppies as the initial population.

$$P(t) = \frac{200}{1 + 19e^{-0.5t}}$$

2. Use your calculator to check the population for one of the next five values of t.

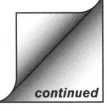

continued

3. Calculate the missing numbers in the table.

4. Graph the data on a piece of graph paper.

5. In what week did the growth rate first start to decrease? How do you know?

6. What is the maximum number of guppies that can live in this tank in the long run? How do you know?

Trip to a Local Field

14

Instructional Purpose

• To implement a transect survey of a field in order to estimate the populations of animals that inhabit the field

Curriculum Alignment  Goal 1 Concept  Goal 2 Content  Goal 3 Process/ Experimental Design ◯ Goal 4 Process/ Reasoning

 Vocabulary

Survey (in this lesson) A detailed inspection of a geographic area for evidence of an organism

Transect To divide by cutting transversely (crosswise) into strips

Quadrat A small, usually rectangular plot of land arranged in a group for close study of plants or animals in an area.

Materials/Resources

• Notepads (one per student)
• Camera and film
• Video camera and blank videotape
• Measuring tapes
• Black construction paper (optional)

• Packets of salt, such as ones given at fast food restaurants (optional)
• Proposal for Estimation Procedure in Field (Handout 14.1)
• Problem Log Questions (Handout 14.2)

Lesson Length

One 60-minute planning session; one 120-minute field activity; one 60-minute debriefing session

Activities: Planning

1. In the video of *The Challenge of the Unknown,* after doing the tag and recapture method, Dr. Sam Gruber did a survey of the shark population in the Bimini Lagoon by transect method. He

flew over the lagoon in a small flying machine called an "ultralight" and counted the sharks that were in each strip as he flew over it. **Ask:**

- What does the word "survey" mean?

- What does the word "transect" mean?

- What are the advantages of estimating by the transect method? Disadvantages?

2. Another way to divide an area for estimating an animal population is to divide it into rectangles called "quadrats" rather than strips as used in the transect method. **Ask:** Why do you think Sam Gruber used transects and not quadrats to divide his map of the Bimini Lagoon?

3. Tell students that they will be going on a field trip to a nearby site to do a transect sampling of animals that are present in the field. Describe the site. **Ask:**

- What animals do you think you will find?

- What factors might affect the numbers of each that are present on the day that you visit?

- How can you make a good estimate of how many of each are in the field?

4. Have students design a survey technique, using **Proposal for Estimation Procedure in Field** (Handout 14.1). (Note: If it is not feasible to go outdoors to do this or if you need to save time, have students do the following activity as a substitute that works as a model of the field activity. Give each group a packet of salt and a piece of black construction paper. The paper is the model field and the salt acts as the model of an organism. Tell students that they should design a method for counting the grains of salt in the packet. Typically this will be done by folding or marking the paper into rectangles (quadrats), sprinkling the salt over the paper, counting the number of grains of salt in the quadrats, and putting the data in a data table.)

Activities: The Field Activity

1. In small groups, have students make a map of the field. Make sure they include a directional marker and scale.

2. Have students record their observations about the field in their notepads. Students should use field guides to identify as many of the plants and animals present as possible. Students should photograph and videotape things of interest.

3. Students should complete the data table that was created on **Proposal for Estimation Procedure in Field** (Handout 14.1).

4. Have students use push pins to mark things of interest on the maps they created.

Activities: Debriefing

1. Hold a debriefing session in which students report on their findings and the whole class analyzes the overall results.

2. Create a chart of group findings. Have students review and discuss the charts. What do they mean?

3. Review the videotape and photographs from the trip. Have students report and discuss their observations. **Ask:**

   • How diverse was the collection of organisms that was present? What does this tell you?

   • Are there any additional things you would want to observe or test if you could go back to the place we visited?

   • How accurate do you think your survey was?

4. Revisit the **Need to Know Board.** Have students reflect about how this field experience contributed to their understanding of the problem situation and a possible resolution.

 ## Notes

1. The size and nature of the field (i.e. mown, overgrown) that you select depend on what kind of organism you would like students to find. If the students are looking for organisms like insects, they can use the field in the center of your school's track or an athletic field if you have one.

2. Before you embark on this field trip, you should try the activity yourself so that you have an idea of what students will encounter. If you find an undesirable organism, such as poison ivy, choose another site.

 ## Homework

Have students complete the **Problem Log Questions** (Handout 14.2) for homework.

 ## Extending Student Learning

Have students research other methods of surveying animal populations.

Assessment

- Proposal for Estimation Procedure in Field: Evaluate the method that students proposed for estimating populations in the field.
- Class discussion: Evaluate how students use their field experience in relationship to the problem situation.

 # Technology Integration

- Students can prepare a digital record of their field experience in the form of a presentation or a portfolio.

Name   Date

Proposal for Estimation Procedure in the Field

1. List the materials that you will need.

2. Write a step-by-step description of what you will do (like a recipe!). List every action you will take during the activity.

continued

3. What data will you be collecting?

4. Design a data table to collect and analyze your information.

Problem Log Questions

1. Suppose you wanted to count the number of crickets in the park across the street from your school. Since you don't have anyone to help you, you divide it into 100 sections, count the number of crickets in one section, and multiply by 100. How confident would you be in your estimate? What assumptions are you making?

2. Do you think that transect sampling would be appropriate for counting deer? How would you implement it?

continued

3. If you went back to the site in a week, would you expect the survey to give different results? What do you think they would be? Explain why.

4. What do you think would happen if you did your deer survey at night instead of in the daytime?

Designing a Survey 15

Instructional Purpose

- To have students design a survey to determine the magnitude of deer-related problems in Kingsford

---

Curriculum Alignment

 **Goal 1** Concept    ● **Goal 2** Content   ○ **Goal 3** Process/ Experimental Design   ● **Goal 4** Process/ Reasoning

---

Materials/Resources

- Maps of the local area
- *Kingsford Gazette* Editorial (Handout 15.1)
- Mayor's Office Memorandum (Handout 15.2)
- Reasoning About a Situation or Event (Handout 15.3)
- Survey Plan (Handout 15.4)
- Problem Log Questions (Handout 15.5)

Lesson Length

60 minutes

 Activities

1. Give the students the **Kingsford Gazette Editorial** (Handout 15.1) and the **Mayor's Office Memorandum** (Handout 15.2). Ask students to read the handouts and update the Need to Know Board.

2. Ask the students to explain what the mayor means by the word "survey." **Ask:**
   - How does the mayor's use of the word relate to Sam Gruber's use of the word? What do you think are the mayor's reasons for wanting a survey? How does she plan to use the results?

3. Divide the students into small groups; have each group brainstorm a list of questions that might be included in the survey. Have each group share its questions with the class; generate a class list of questions.

4. Ask the students if they think many people would volunteer to answer a long list of questions

like the one they have made, or if people are more likely to complete a survey form if it is short. Ask the students how many pages they would be willing to complete if they were responding to a survey and the amount of time that they would be willing to spend on one. Ask students if they think there are any groups of people who are more likely to complete the survey (gender, age, SES, etc.). Based on the results of this discussion, divide the students into small groups again and have them make a list of the most important questions to ask on the survey form, with the most important question first, the second most important second, and so on. Limit them to approximately 10 questions.

5. Discuss the groups' lists as a class. Have the students develop a survey form that everyone is willing to use.

6. Review the reasoning vocabulary introduced in Lesson 4. Use **Reasoning About a Situation or Event** (Handout 15.3) to help students determine the stakeholder groups that may be interested in the community's problem and therefore might be surveyed.

7. Ask the students which stakeholders in the community of Kingsford should be asked to complete a survey form. If they suggest that everyone should, remind them that the survey is supposed to be quick and inexpensive. Remind the students that they already know something about sampling based on their earlier experience with using sampling to count animal populations. Ask them if some of the techniques used in that lesson would be useful here. (Students should suggest giving the survey to a sample of the people in the community and using the results to predict what the whole community's deer experience has been.)

8. Ask the students how they could develop a representative sample of people in the community. Ask them whether publishing the survey in the *Gazette* and asking readers to send it to the Mayor's office would result in an unbiased sample. What other ways to distribute the survey can they think of? Which of these would generate a fair sample of the community's opinions on deer?

9. Divide the students into small groups and have them develop a plan for the survey that would give the mayor the information she needs. Have each group create a sampling plan, then describe it on the **Survey Plan** (Handout 15.4).

 **Homework**

Have students complete the **Problem Log Questions** (Handout 15.5) for homework.

Extending Student Learning

1. Have students contact a local government office that does mail or telephone surveys to find out how they design and implement them.

2. Students may use their questions to survey their school or neighborhoods, if their area has deer. The students could also do a survey of their school/neighborhoods to determine the extent of the local Lyme disease problem, if there is any. If they find information, they can use push-pins to mark trouble spots on the map. Use the problem log questions in Handout 13.4 as a guide for this activity.

3. Have students research the history of deer populations in the United States, in your state, and in your local area both over long periods of time (from pre-colonial times until the present) and during recent years. (The November 1995 issue of *Virginia Wildlife* has excellent information about deer populations in the state of Virginia.)

 ## Assessment

- Question lists: Evaluate student reasoning in determining the types of questions to include that will assist in resolving the problem.

Name   Date

Kingsford Gazette Editorial

From the Editor's Desk
by Virginia Olin

It is the season to plant new trees and flowers in anticipation of the shade and beauty they will bring. Garden centers overflow with brightly colored annuals. New trees, looking like bundles of sticks, await a place of honor in their new homes. Every sunny weekend morning you can find gardeners on their hands and knees, tenderly encouraging the growth of their young plants. Like many of my neighbors, I too planted geraniums in the flowerpots on the front porch. My work, which was to be appreciated by every passerby, was all for naught. The very first evening after my flowers were planted, a doe sauntered out of the woods behind my house, nimbly climbed my front steps, and ate my geraniums for dinner. If my garden were the only one raided by deer, I would be the last to complain. I feel, though, that I speak for many of my fellow gardeners in calling for the mayor to do something about this terrible deer overpopulation problem. Everyone I talk to has a similar story. How long must we wait for something effective to be done?

Name   Date

Mayor's Office Memorandum

MEMO

To: Deputy Mayor Marie Barnes

From: Jane White, Mayor

Marie,

If I thought Virginia Olin's geraniums were the only casualties of deer browsing, I would do my best to ignore this—but I've been getting a lot of complaints since that first letter from Ira Stanley got published in the *Gazette*. I need to know how much of a problem we really have with deer here in Kingsford. Are there many bad feelings in the community, or are Virginia and Ira and their buddies the only people who really care?

Let's do a quick, cheap survey and find out—I'd like a plan for one by Monday.

Jane

Name _____ Date _____

Reasoning About a Situation or Event

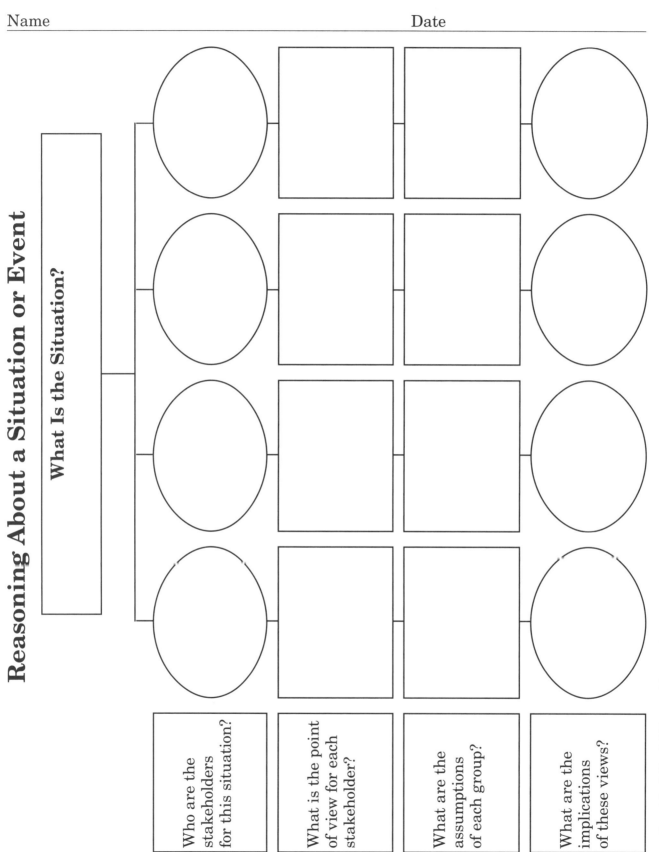

What Is the Situation?

Who are the stakeholders for this situation?

What is the point of view for each stakeholder?

What are the assumptions of each group?

What are the implications of these views?

Name _____   Date _____

Survey Plan

1. Which stakeholders will you ask to complete the survey?

2. Why did you choose the people you chose?

continued

3. How have you made sure that your sample represents all of different neighborhoods in town?

4. What percentage of people do you predict will respond to your survey? Why do you think that? Will there be enough data to tell the mayor what she needs to know?

5. How could you find out what the appropriate rate of return on a survey is?

Name _____   Date _____

Problem Log Questions

Post a street map or road map on a bulletin board and use colored push-pins to identify places where there have been complaints about deer, accidents involving deer, etc.

1. Describe the places where deer appear or congregate. What makes these places attractive to deer?

2. What times of day are deer usually seen in these places?

continued

3. Does weather have any effect on when and where they are seen?

4. Are there seasonal variations in the patterns of deer appearances in neighborhoods?

5. Have deer populations really gotten bigger recently or have people just reported more problems because there are more people?

Visit from a Wildlife Biologist 16

Instructional Purpose

- To learn about the extent of the deer problem in the local area
- To learn about wildlife management in the local area
- To learn how to use experts as information resources

Curriculum Alignment   **Goal 1** Concept   **Goal 2** Content   **Goal 3** Process/ Experimental Design   **Goal 4** Process/ Reasoning

 ## Materials/Resources

- Chart paper
- Markers
- Audiovisual equipment for guest speaker
- Visitor Planning Questions (Handout 16.1)

Lesson Length

Three 60-minute sessions (This lesson will be composed of three sessions: **Planning for the Guest Speaker, The Guest Speaker's Presentation,** and **Debriefing.**)

 ## Activities: Planning for the Guest Speaker

1. Have students brainstorm questions to ask the guest speaker. Use the **Need to Know Board** as a basis for formulating questions. During the class discussion, help students sort questions into the most and least important questions to be addressed.

2. Students should also be guided to think about the best way to phrase the questions. Are they specific enough? Are they too specific?

3. As students are brainstorming questions, **Ask:**
   - What information do we want to know?
   - What information will the guest speaker be most qualified to give?
   - What do we want to know by the time the guest speaker leaves?

- What facts do we want to get from this person?
- What opinions would be interesting to learn?
- Which of these questions are most important?
- How can we get an idea of this person's perspective on this kind of situation?
- Do you think this person will have a bias? What would it be? How can we find out?

4. Record group questions on a master question chart.

5. Students can then add any of their own questions to the individual **Visitor Planning Questions** (Handout 16.1).

 ## Activities: The Guest Speaker's Presentation

1. Guest Speaker: The guest provides his/her information regarding the area of his/her expertise.

2. Students take notes and ask their questions during and after the presentation.

3. Students should also be prepared to share with the guest speaker background on the problem and their decisions to date.

 ## Activities: Debriefing

1. As a follow-up to the guest speaker, teacher and students should review the **Need to Know Board,** removing questions which have been answered and adding new issues, if necessary.

2. Teachers and students should discuss the potential bias in the information provided by the guest speaker and the possible effects of that bias on the validity of the information.

3. During the debriefing session, **Ask:**
   - What information did we learn from the guest speaker?
   - How does the new information affect our thinking about the problem?
   - Do we need to reorganize or revise our approach to the problem?
   - Did this person reveal a particular bias? If so, what?
   - Where can we go to get another perspective so that we have a balanced report of information?

4. Revise the **Need to Know Board** if necessary.

Problem Log

Have students respond to this prompt in their **Problem Logs:** What was the guest speaker's point of view regarding the problem? Did this reflect a particular bias? If so, what was the bias?

 Notes

1. The planning portion of this lesson may be conducted well in advance of the guest speaker's visit.

2. If the expert comes to the classroom, all students can participate. This format can also be used by small groups who need to interview an outside expert outside of class; afterwards, they can report any new information to the class.

 Homework

Have students write thank you letters to the guest speaker. In their letters, they should describe which information was especially helpful for their problem exploration.

 Extending Student Learning

Have students research various jobs that involve work with wildlife and wildlife management.

 Assessment

- **Problem Logs:** Assess student understanding of point of view and bias.

 Technology Integration

- If a speaker cannot come into the classroom, the interview can be conducted online via email or messaging/chat software.

- If the guest speaker is willing to do so, he/she may act as a tele-mentor through e-mail exchanges during the remainder of the unit.

Name _____   Date

Visitor Planning Questions

Name of Visitor _____

What is the visitor's role regarding wildlife management?

_____

_____

_____

Why is this visitor coming to see us?

_____

_____

Why is this visitor important to us? How might he/she assist with our problem resolution?

_____

_____

What would you like to tell our visitor about our problem?

_____

_____

What questions do you want to ask the visitor?

_____

_____

Humans vs. Deer 17

Instructional Purpose

- To discuss issues regarding the relationship between humans and wildlife
- To debate the ethics of hunting wildlife

Curriculum Alignment   Goal 1 *Concept*   Goal 2 *Content*   Goal 3 *Process/ Experimental Design*   Goal 4 *Process/ Reasoning*

 Vocabulary

Biological carrying capacity The population that the environment can support for an extended period of time without damage to the habitat. Usually "carrying capacity" refers to biological carrying capacity.

Cultural carrying capacity The maximum animal population that a human population is willing to tolerate in a given area

 Materials/Resources

- *Noah's Garden: Restoring the Ecology of Our Own Backyards* by Sara B. Stein
- Article from *The Kingsford Gazette* (Handout 17.1)

- The Pros and Cons of Hunting Deer (Handout 17.2)
- Debate Format (Handout 17.3)
- Developing Your Own Point of View (Handout 15.4)

Lesson Length

Two 60-minute sessions

 Activities: Considering Different Points of View

1. Distribute **Article from *The Kingsford Gazette*** (Handout 17.1). Have students update the **Need to Know Board** given this information.

2. Review the reasoning vocabulary introduced in

173

Lesson 4. Have students discuss stakeholders and their points of view in this problem situation.

3. Read aloud or have students read the first eight pages of Chapter 5 of Sara Stein's book, *Noah's Garden: Restoring the Ecology of Our Own Backyards.* They should write a paragraph in response to the question, "What does the author mean by the chapter title 'Who Gets to Stay Aboard the Ark?'" This task may be completed as homework before this class session.

4. Engage students in a discussion of their responses to the question posed in #3.

5. Extend the discussion. **Ask:**

   • What assumptions has the "famous gardener" on the second page of the chapter made about snakes and their habitat? Are they correct assumptions?

   • Where does Sara Stein live? How much of what she says applies to your area of the country? Explain.

   • What information about deer did you find in this chapter? Add it to the **Need to Know Board.**

   • How has suburban sprawl helped the deer?

   • What are the options for controlling deer that are mentioned in this chapter?

   • How credible a source of information do you think Sara Stein is? Should you believe what she says?

   • What is the difference between "carrying capacity" and "cultural carrying capacity" that is mentioned on the seventh page of the chapter?

   • Do you think that Sara Stein is sympathetic to the sport of hunting deer? What evidence can you find in the text that supports this point of view?

6. In their **Problem Logs,** have students list the pros and cons of hunting deer; they should use the format of **The Pros and Cons of Hunting Deer** (Handout 17.2) in preparation for the debate in the second part of this lesson. They may need prompting to include arguments related to economics, environmental factors, health, safety, recreation, and cultural traditions.

 ## Activities: Preparing to Debate

1. Distribute **Debate Format** (Handout 17.3) and use it to teach the skill of debating to students.

2. Students will debate the following topic:

   Resolved: That hunting is immoral and unethical in the United States in the 2000's.

   You may use **Article from *The Kingsford Gazette*** (Handout 17.1) to show students how some stakeholders view the issue.

3. Assign teams to be in favor of or against the resolution. Allow students time to plan their arguments.

4. Using **Debate Format** (Handout 17.3), have the teams debate. One team debates, and the other team scores and marks the argumentation points made. Then reverse the roles.

5. Discuss the debate with the whole class. Discuss how the issues discussed in the debate could be helpful when considering a way to resolve the problem.

6. Distribute **Developing Your Own Point of View** (Handout 17.4). Students will write a letter arguing their own points of view on hunting to the editor of the newspaper.

 ## Notes

1. Teachers may want to refer to the following article for more information on debate: "Debating with Gifted Fifth and Sixth Graders—Telling it Like It Was, Is, and Could Be," by Barbara Swicord, in *Gifted Child Quarterly, 28* (3), Summer 1984, pp. 127–129.

2. Contact the state game department to get the following resources:
   - A video about the urban deer problem: "Whitetails at the Crossroads," produced under a grant from the U. S. Fish and Wildlife Service in Cooperation with the Northeast Deer Technical Committee. It runs for 28 minutes.
   - State and local laws governing hunting.

 ## Extending Student Learning

Have students complete the following extension activities:

1. Read the remainder of Chapter 5 of *Noah's Garden: Restoring the Ecology of Our Own Backyards.*

2. Read *My Side of the Mountain* by Jean Craighead George. The story chronicles the life of a teenager who runs away from New York City to live off the land in the Catskill Mountains. (This novel will be easy for middle school students to read but will provide some interesting insights into the utilitarian side of hunting.)

3. Debate the issue of hunting buffalo that wander outside the boundaries of Yellowstone Park. Is this issue different from the deer issue? In what ways?

 ## Assessment

- Evaluate students' reasoning as they participate in the debate.

Article from *The Kingsford Gazette,* Monday, June 16, 2006

From the Editor's Desk
Virginia Olin, Editor

Twenty years ago, Kingsford was a sleepy coastal town. The age-old traditions of country living gave a rich glow to our existence.

Now Kingsford is suffering unparalleled growth and change. New housing developments are springing up like toadstools after a rain. Suburban sprawl has obliterated fields and forest, forcing wildlife into close proximity to people.

The new suburbanites have come to enjoy Kingsford's relaxed charm, but they have brought attitudes with them that make Kingsford a divided community. The conflict between old-timers who remember killing their first buck at age 12 and newcomers raised on the movie "Bambi" has polarized the community.

The problems we have with deer make it clear that something must be done. Hunting is a venerable American tradition, one that Kingsford residents have long enjoyed. We think a well-controlled annual deer hunt would be a solid approach to keeping Bambi at bay.

Name _____   Date _____

The Pros and Cons of Hunting Deer

Pros	Cons

Name _____   Date _____

Debate Format

What is a debate?

Debate is a special type of argument in which two or more speakers present opposing propositions in an attempt to win the audience to their sides. The teams are not concerned with convincing each other. The purpose is to try to alter audience thinking by presenting the issues using supporting evidence.

Why debate?

Debate helps you to:

1. Analyze problems.

2. Reinforce statements with proof.

3. Express your ideas clearly.

4. Gain confidence.

5. Think quickly.

6. Understanding alternate viewpoints upon reflection.

7. Engage in a logical thought process.

8. Speak in public.

9. Persuade others to a different point of view.

What are the rules of debating?

Debates begin with a proposed solution to a problem. The proposal should begin with the word RESOLVED. Examples:

- Resolved: That the United States should abolish the electoral college and elect the President by popular vote.
- Resolved: That elementary students should wear school uniforms.

1. The same number of persons speak on each opposing side.

2. Begin with careful analysis of the subject to be debated by both teams. Each member should know as much about the opponent's arguments as she does about her own position. (Note to teachers: "Switch side" debating requires this as teams do not know which side they'll be taking until the debate begins.)

3. Decide which arguments are closely related and worthy of being included, and which are irrelevant and should be excluded.

continued

4. List the main issues for each side.

5. Find evidence (facts, examples, statistics, testimony) that will prove the issue true and false.

6. Be prepared to answer the arguments made by the other team. This response is called a REBUTTAL and presents opposing evidence or points out flaws in the argument.

What is the format for a debate?

Suggested Procedure:

> First affirmative speech—5 minutes
> First negative speech—5 minutes
> Second negative rebuttal—2 minutes
> Second affirmative rebuttal—2 minutes

The debate always begins and ends with the affirmative team.

One way of scoring can be done by giving:

1. One point for an argument or

2. Two points for an argument with proof.

Name   Date

Developing Your Own Point of View

Now that you have heard the debate on the issue of hunting, develop your own point of view about hunting a particular kind of animal in your area. Write a letter to the editor of *The Kingsford Gazette* stating your point of view and supporting it with reasons.

Problem Resolution

18

Instructional Purpose

• To have students develop proposals for solutions to the deer problem
• To have students present their solutions to the problem to an audience

 **Curriculum Alignment** ◯ **Goal 1** Concept  ● **Goal 2** Content ● **Goal 3** Process/ Experimental Design ◯ **Goal 4** Process/ Reasoning

 ## Materials/Resources

- Role sheets for adult volunteers (teacher-made)
- Question sheets for adult volunteers (teacher-made)
- Editorial from *The Kingsford Gazette* (Handout 18.1)
- Memo from the Mayor (Handout 18.2)
- Analysis of All Possible Solutions (Handout 18.3)
- Group Solution Sheet (Handout 18.4)

Lesson Length

Three 60-minute sessions (This lesson will be composed of three sessions: Developing Solutions, City Council Meeting, and Debriefing.)

 ## Activities: Developing Solutions

1. Give students the **Editorial from *The Kingsford Gazette*** (Handout 18.1) and ask them to read it. **Ask:** Does the information in the editorial change the problem? If so, how?

2. Distribute **Memo from the Mayor** (Handout 18.2) and discuss the need to develop possible solutions. Divide the students into groups to discuss and have them research solutions to the problems listed by the mayor. Distribute **Analysis of All Possible Solutions** (Handout 18.3) and **Group Solution Sheet** (Handout 18.4) to provide scaffolding for their thinking.

3. As a class, discuss each of the group's possible solutions. For each of the sub-problems, have students rate the efficacy of the proposed solutions. Ask the students to rank the proposals and explain the reasons for their ranking.

 ## Activities: City Council Meeting

1. Before the city council meeting, find adult volunteers to play the parts of the mayor (Jane White) and various concerned stakeholders. Develop role sheets to acquaint each volunteer with his/her role.

2. At the meeting, have each group of students present the class solutions to the deer problem. One group could present alternative solutions for garden damage; one group could present solutions for vehicle/deer problems; one group could talk about managing the deer population itself; and the last group could talk about Lyme disease prevention. Provide the volunteers with appropriate questions to ask the members of each group. Have the mayor ask each group of students for their recommendations, and for a justification of their recommendations.

 ## Activities: Debriefing

1. During a session when the adults have left, have a debriefing session about the presentations to the city council. **Ask:**
   - Did anyone else's arguments change your own point of view?
   - What questions are still unanswered?
   - What have you learned from this?

 ## Extending Student Learning

For each solution, have the students write a plan for a pilot test of its efficacy. They can use the experimental design forms to plan their test.

 ## Assessment

- Group solution sheets: Evaluate student reasoning in developing possible solutions.
- Presentations to city council: Evaluate oral presentation skills.

Editorial from *The Kingsford Gazette*

From the Editor's Desk
by Virginia Olin

Tuesday's city council meeting promises to at last force our mayor to come up with a way to deal with the current crisis in wildlife management. We understand that her staff has been working frantically to come up with proposals for controlling the deer population and its threats to our gardens, cars, and health. We hope that these proposals will be detailed, sensible, and acceptable to the people of this community. The time for action is long past—just ask the parents of little Melanie Roberts, who is still on antibiotic therapy for her Lyme disease-inflicted arthritis. We urge citizens to attend this meeting and tell the mayor what they think of her ideas. The future health and happiness of the community are at stake!

Memo from the Mayor

MEMO

To: Deputy Mayor Marie Barnes

From: Jane White, Mayor

Marie,

How do you like Virginia's latest editorial? Maybe you had better not answer that. . .

We need to get a short list of options together to discuss at the city council meeting. For each option, I need a list of pros and cons. I need to know what the questions are likely to be for each one so I'll be prepared to answer them at the meeting. Could you also give me an idea of which option you think is the best, and why?

Thanks,

Jane

Name _____   Date _____

Analysis of All Possible Solutions

1. What are your criteria for a good solution?

2. Make a list of all possible solutions to the problem. Rate each solution on each of the criteria.

Name _____   Date _____

Group Solution Sheet

1. What is your solution to the problem that you are trying to solve?

2. How does your solution attempt to solve the problem?

3. Which of the sub-problems does your solution solve?

continued

4. What groups of people will not like your solution and why?

5. How expensive will your solution be? Explain how you made the estimate. Compare this to the cost of other solutions. Where will the money come from to pay for it?

6. What aspects of the problem does your proposed solution fail to address?

Post-Assessments

19

Instructional Purpose

- To assess student understanding of experimental design
- To assess student understanding of population biology
- To assess student understanding of models

Curriculum Alignment Goal 1 _Concept_ Goal 2 _Content_  Goal 3 _Process/ Experimental Design_ ●Goal 4 _Process/ Reasoning_

 Materials/Resources

- Experimental Design Post-Assessment (Handout 19.1)
- Experimental Design Rubric (Teacher Resource 1)
- Content/Concept Post-Assessment (Handout 19.2)
- Content/Concept Post-Assessment Scoring Guide (Teacher Resource 2)
- Process Post-Assessment (Handout 19.3)
- Process Post-Assessment Scoring Guide (Teacher Resource 3)

Lesson Length

180 minutes

 Activities

1. Congratulate students on their success with the unit and debrief about their experiences. Have students share what they learned.

2. Distribute the **Experimental Design Post-Assessment** (Handout 19.1) and have students complete it individually. Collect and score the assessments using the **Experimental Design Rubric** (Teacher Resource 1).

3. Distribute the **Content/Concept Post-Assessment** (Handout 19.2) and have students complete it individually. Score **Content/Concept**

Post-Assessment using **Content/Concept Post-Assessment Scoring Guide** (Teacher Resource 2).

4. Distribute the **Process Post-Assessment** (Handout 19.3) and have students complete it individually. Collect and score the assessments using the **Process Post-Assessment Scoring Guide** (Teacher Resource 3).

5. Have students compare their experimental design pre-assessment responses to their post-assessment responses. Have them reflect about what they have learned and how they have grown as scientists throughout the course of the unit.

 ## Problem Logs

Have students respond to one of the following prompts:

- I want to learn more about these things in science. . .
- Studying science can help me in the following ways . . .
- When I want to know more about a topic in science, I. . .

 ## Note

The post-assessments in this lesson are lengthy; you may wish to have students complete them over several days. However, it is important to emphasize that students should not share their ideas with each other until all three assessments are complete.

 ## Assessment

- Experimental Design Post-Assessments
- Content/Concept Post-Assessment
- Process Post-Assessment

Name _____   Date _____

Experimental Design Post-Assessment
(45 minutes)

Construct a fair test of the following question: Which brand of flashlight battery (Brand A, Brand B, or Brand C) has the longest life?

Describe in detail how you would test this question. Be as scientific as you can as you write about your test. Write the steps you would take to find out which of the three brands lasts the longest (that is, has the longest life).

Adapted from Fowler, M. (1990). The diet cola test. *Science Scope, 13(4)*, 32–34.

Teacher Resource 1: Experimental Design Rubric

Criteria	Strong Evidence 2	Some Evidence 1	No Evidence 0	Pre	Post
States **PROBLEM** or **QUESTION**.	Clearly states the problem or question to be addressed.	Somewhat states the problem or question to be addressed.	Does not state the problem or question to be addressed.		
Generates a **PREDICTION** and/or **HYPOTHESIS**.	Clearly generates a prediction or hypothesis appropriate to the experiment.	Somewhat generates a prediction or hypothesis appropriate to the experiment.	Does not generate a prediction or hypothesis.		
Lists experiment steps.	Clearly & concisely lists four or more steps as appropriate for the experiment design.	Clearly & concisely lists one to three steps as appropriate for the experiment design.	Does not generate experiment steps.		
Arranges steps in **SEQUENTIAL** order.	Lists experiment steps in sequential order.	Generally lists experiment steps in sequential order.	Does not list experiment steps in a logical order.		
Lists **MATERIALS** needed.	Provides an inclusive and appropriate list of materials.	Provides a partial list of materials needed.	Does not provide a list of materials needed.		
Plans to **REPEAT TESTING** and tells reason.	Clearly states a plan to conduct multiple trials, providing reasoning.	Clearly states a plan to conduct multiple trials.	Does not state plan or reason to repeat testing.		
DEFINES the terms of the experiment.	Correctly defines all relevant terms of the experiment.	Correctly defines some of the relevant terms of the experiment.	Does not define terms, or defines terms incorrectly.		
Plans to **MEASURE**.	Clearly identifies plan to measure data.	Provides some evidence of planning to measure data.	Does not identify plan to measure data.		
Plans **DATA COLLECTION**.	Clearly states plan for data collection, including note-taking, the creation of graphs or tables, etc.	States a partial plan for data collection.	Does not identify a plan for data collection.		
States plan for **INTERPRETING DATA**.	Clearly states plan for interpreting data by comparing data, looking for patterns and reviewing previously known information.	States a partial plan for interpreting data.	Does not state plan for interpreting data.		
States plan for drawing a **CONCLUSION BASED ON DATA**.	Clearly states plan for drawing conclusions based on data.	States a partial plan for drawing conclusions based on data.	Does not state plan for drawing conclusions.		
			TOTAL SCORE:		

Adapted from Fowler, M. (1990). The diet cola test. *Science Scope, 13(4)*, 32–34.

Name _____   Date _____

Content/Concept Post-Assessment
(90 minutes)

1. Describe the role of deer in the spread of Lyme disease.

2. If there were no deer, would Lyme disease still be able to spread? Why or why not?

3. List three possible ways to control deer populations. For each method, list one advantage and one disadvantage.

continued

4. Deer populations in the Eastern United States have been increasing in size since World War II. Based on what you know about the ecological niche occupied by deer, explain this increase.

5. For the following graph:

   a. Identify the curve that shows exponential population growth.

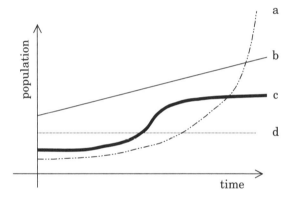

   b. Identify the curve that shows a steady state population.

   c. Explain which curve resembles the curve for the human population since 1800.

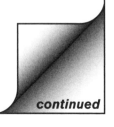

continued

d. Which curve is most likely to represent a species that will be able to continue its existence over time without encountering an ecological limit? Explain.

e. Which curve is most likely to represent a population that is constrained by its carrying capacity?

f. Is the graph in this problem a physical, conceptual, or mathematical model? Explain.

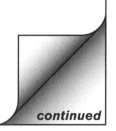

continued

6. List three precautions people can take to avoid getting Lyme disease. Explain why each of the precautions you have listed is necessary.

7. Give an example of each:

   a. a physical model

   b. a conceptual model

   c. a mathematical model

Content/Concept Post-Assessment Scoring Guide
(Teacher Resource 2)

1. Describe the role of deer in the spread of Lyme disease.

   Answer: White-tailed deer are parasitized by deer ticks. Deer ticks are frequently infected with Borrelia burdorfi, *the organism that causes Lyme disease. When an infected tick bites an uninfected deer, the deer can become infected. Uninfected ticks that bite the infected deer then become infected themselves. Because deer can travel great distances, infected ticks can be spread into new territory. When infected ticks drop off the deer, they can then move onto humans or household pets. A bite from an infected tick will cause a person to become infected and to develop Lyme disease. Deer thus have two major roles in the spread of the disease: they facilitate the infection of previously uninfected ticks, and they allow infected ticks to move from place to place.*

   (Accept any answer that mentions the two roles described above.)

2. If there were no deer, would Lyme disease still be able to spread? Why or why not?

   Answer: Yes, Lyme disease will still be able to spread because there are other wild animal hosts such as rodents.

3. List three possible ways to control deer populations. For each method, list one advantage and one disadvantage.

   Accept any reasonable answers. Here are three examples.
   - *Hunting: Advantage is efficacy; disadvantage is public attitude.*
   - *Starvation and disease: Advantage is that it is natural; disadvantage is public reaction and effects on the ecosystem of desperate attempts of deer to find food.*

4. Deer populations in the Eastern United States have been increasing in size since World War II. Based on what you know about the ecological niche occupied by deer, explain this increase.

   Deer prefer the forest edge. Suburbia and farms provide an edge habitat for deer. Since WWII, the amount of land devoted to suburban development has increased tremendously. This, coupled with restrictions on hunting and a lack of natural predators, has allowed deer populations to increase.

   Accept all answers that mention the increase in available edge habitat and limits on human and animal predation.)

continued

5. For the following graph:

   a. Identify the curve that shows exponential population growth.

      Answer: a

   b. Identify the curve that shows a steady state population.

      Answer: d

   c. Explain which curve resembles the curve for the human population since 1800.

      Answer: a

   d. Which curve is most likely to represent a species that will be able to continue its existence over time without encountering an ecological limit? Explain.

      Answer: d. The steady state population is less likely to encounter an ecological limit because it is not in a position to outgrow its resources (assuming that there are no other changes in the ecosystem).

   e. Which curve is most likely to represent a population that is constrained by its carrying capacity?

      Answer: c. This population tapers off to approach a steady state when it nears its carrying capacity.

   f. Is the graph in this problem a physical, conceptual, or mathematical model? Explain.

      The graph is a physical model of the size of the population over time. If the curves are used to extrapolate current population trends to the future, then the graph is being used as a mathematical model as the shape of the curve, a mathematical entity, is being used as a predictive tool.

6. List three precautions people can take to avoid getting Lyme disease.

   - *People can avoid areas known to be infested by deer ticks, such as woodlands and high grass.*
   - *People can wear protective clothing, such as long sleeved shirts and long pants, to prevent the attachment of deer ticks to bare skin.*
   - *People can use insect repellent to deter ticks.*

      (Accept all reasonable answers.)

continued

7. Give an example of each: (accept all reasonable answers)

   a. a physical model—*a scale model of a building*

   b. a conceptual model—*the idea of a light bulb could be a model for a brain getting an idea; it lights up when the idea occurs in the brain.*

   c. a mathematical model—*an equation such as $D = R \times T$*

Name _____   Date _____

Process Post-Assessment
(45 minutes)

In a *New York Times* article ("Roissy Journal: Invincible Rabbit Army Besieges a Paris Airport," Thursday, August 8, 1996, page A4), the journalist, Craig Whitney, described a problem faced by the managers of Charles de Gaulle Airport. The airport is overrun with rabbits. The gamekeeper Jean Valissant estimates that there are 50,000 of them living in the 5,000 acres of fenced-in grass and short cornfields that line the two airport runways.

1. One of Jean Valissant's tasks is to estimate the number of rabbits present at the airport.

   a. If you were the gamekeeper at the airport, what equipment and resources would you need to make your estimate?

   b. What procedure would you use to make your estimate? Describe what you would do and how you would use the results to estimate the number of adult rabbits present at the airport.

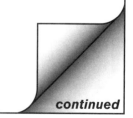

continued

c. How could you check to be sure that your estimate was accurate? Describe the procedure you would use and how you would use your results to check your estimate.

d. List at least two models that you would use during your estimation procedure and explain whether they are physical models, conceptual models, or mathematical models. What limitations do each of these models have?

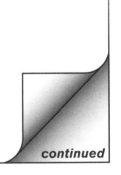

continued

2. Infectious myxomatosis is a disease that is fatal to rabbits. It has been used in Australia to control rabbit populations. Suppose an outbreak of this virus killed all but one male and one female rabbit at the airport, and that these two rabbits were able to find each other and start a family. Rabbits usually have two litters a year, with five babies per litter.

   a. Assuming no rabbits die once the viral outbreak is over, sketch a graph showing the rabbit population every year for five years after the viral outbreak. Numbers don't matter—it's the shape of the curve that matters.

   b. What kind of growth pattern does this rabbit population show?

   c. Would a one-time use of myxomatosis virus solve the rabbit problem at the airport? Explain.

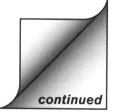

continued

3. One rabbit population control measure used at the airport is the capture of rabbits by men equipped with nets and ferrets, animals which can chase rabbits out of their burrows. Mr. Valissant said that 3,775 rabbits were captured and removed from the airport last year.

a. If half of the rabbits at the airport are female and each female rabbit has 10 babies per year, estimate how long it would take for the rabbits that were removed from the airport to be replaced by new baby rabbits? Explain your reasoning.

b. Do you think that capture of rabbits by men with nets and ferrets is an effective way to deal with the problem? Why or why not?

Process Post-Assessment Scoring Guide
(Teacher Resource 3)

In a recent *New York Times* article ("Roissy Journal: Invincible Rabbit Army Besieges a Paris Airport," Thursday, August 8, 1996, page A4), the journalist, Craig Whitney, described a problem faced by the managers of Charles de Gaulle Airport. The airport is overrun with rabbits. The gamekeeper Jean Valissant estimates that there are 50,000 of them living in the 5,000 acres of fenced-in grass and short cornfields that line the two airport runways.

1. One of Jean Valissant's tasks is to estimate the number of adult rabbits present at the airport.

   a. If you were the gamekeeper at the airport, what equipment and resources would you need to make your estimate?

   You would need nets or traps to catch them, tags or some other way of marking rabbits that had been caught, protective gear (such as gloves) to protect people handling these wild animals, some way to manage and analyze the data (everything from paper and pencils to a computer) and people to do the work.

   (Accept any reasonable answer.)

   b. What procedure would you use to make your estimate? Describe what you would do and how you would use the results to estimate the number of adult rabbits present at the airport.

   I would use the tag and recapture method to estimate the rabbit population. On day one, my helpers and I would set humane traps loaded with rabbit bait at random points around the airport. We would examine the traps the next day, tag any rabbits we found, and set them free. A few days later, we should set humane traps loaded with rabbit bait in a new set of random points around the airport (to avoid biasing our results in favor of rabbits who were in the habit of hopping past the previous trap sites.) We would examine the traps the next day and determine what fraction of the rabbits were tagged. We would then use proportional reasoning to estimate the total size of the rabbit population.

   (Accept any reasonable answer.)

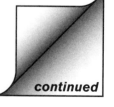

continued

c. How could you check to be sure that your estimate was accurate? Describe the procedure you would use and how you would use your results to check your estimate.

My helpers and I would capture, tag, and count all of the adult rabbits in several small areas of the airport and determine the average number of rabbits per unit area. We would then multiply this number by the total area of the airport to estimate the total number of adult rabbits at the airport. If this estimate and the tag and recapture estimate were similar, we could feel confident that we had a good estimate of the number of adult rabbits at the airport.

(Accept any reasonable answer as long as it uses a different method of estimation than the one mentioned in the student's answer to part b.)

d. List at least two models that you would use during your estimation procedure and explain whether they are physical models, conceptual models, or mathematical models. What limitations do each of these models have?

- *I would need a map of the airport to choose sites for traps and for intensive rabbit counting. This is a physical model depending on the kind of map I was using, different kinds of information would be missing (there's no way to tell from an aerial photo whether there are snakes in the grass where I'm planning to work, for example) so a trip to the potential trapping sites would be necessary to confirm their usefulness.*

- *My estimates would be based on mathematical equations, which are mathematical models. Which equations to use is dependent on my assumptions about the rabbits and the airport (for example, in the tag and recapture method I am assuming that rabbits that had been caught once would be equally likely to be caught a second time and that the tagged rabbits will be evenly distributed across the airport when it's time to recapture them). These assumptions may or may not be valid; the validity of the assumptions limits the usefulness of the model.*

(Accept any other reasonable answers.)

continued

2. Infectious myxomatosis is a disease that is fatal to rabbits. It has been used in Australia to control rabbit populations. Suppose an outbreak of this virus killed all but one male and one female rabbit at the airport, and that these two rabbits were able to find each other and start a family. Rabbits usually have two litters a year, with five babies per litter.

    a. Assuming no rabbits die once the viral outbreak is over, sketch a graph showing the rabbit population every year for five years after the viral outbreak. Numbers don't matter—it's the shape of the curve that matters.

    Answer: A curve showing exponential growth should be sketched by the student.

    b. What kind of growth pattern does this rabbit population show?

    Answer: This rabbit population is growing exponentially.

    c. Would a one-time use of myxomatosis virus solve the rabbit problem at the airport. Explain.

    Answer: No, because if even one breeding pair escaped, the airport would be quickly repopulated by rabbits.

3. One rabbit population control measure used at the airport is the capture of rabbits by men equipped with nets and ferrets, animals which can chase rabbits out of their burrows. Mr. Valissant said that 3,775 rabbits were captured and removed from the airport last year.

    a. If half of the rabbits at the airport are female and each female rabbit has 10 babies per year, estimate how long it would take for the rabbits that were removed from the airport to be replaced by new baby rabbits? Explain your reasoning.

    Answer: If half of the rabbits at the airport were female, and each female produced ten babies per year, then there would be 250,000 baby rabbits born per year. Divide this by 365 and you find that there are 685 baby rabbits born each day. 3,775 divided by 685 is 5.5. Thus, Mr.. Valissant's captured rabbits would be replaced in five and a half days.

    b. Do you think that capture of rabbits by men with nets and ferrets is an effective way to deal with the problem? Why or why not?

    Answer: It is not an effective way to deal with the problem, because the rabbit population grows much more quickly than Mr. Valissant's helpers can reduce it.

3

Implementation Guidelines

Implementation Guidelines

The following pages provide guidelines for teachers to implement this unit effectively in classrooms, including some design and logistical discussions. Copies and explanations of the teaching models are also included.

1. Implementation Considerations

Target Population

This unit was designed to serve the learning needs of highly able students in the middle school grades (6 to 8). Lessons have been piloted both in classes for the gifted and in heterogeneous settings, with teachers modifying some reading selections and activities for use with some students as appropriate.

Alignment of the Unit with Standards

The unit was designed to align with the grade six standards of the *Standards of Learning for Virginia Public Schools* (Commonwealth of Virginia Department of Education, 2001). Although the content of the unit reflects content sometimes addressed in later grades, unit activities are targeted toward younger students. The unit also aligns with the standards set forth by the *National Science Education Standards* and the *Benchmarks of Scientific Literacy* with regard to process, content, and concept elements, including emphases on logical thinking, data collection and interpretation, and experimental design. The unit also supports student learning in other areas, particularly in the language arts area of persuasive writing.

Schedule for Unit Implementation

Recognizing the limited time often allotted to science in the school schedule, lessons incorporated in this unit might require more than one class period to implement. Teacher judgment is recommended as to where to split the lessons.

Use of Technology

Internet access and other technological tools will provide support for unit implementation and enhance the experience for students and teachers. A number of Internet sites are listed in *Part 4, Appendix B, Suggested Mini-Lessons*. In addition, the Internet is a useful resource for completing the unit activities. Beyond this, computers should be utilized to support student word-processing skills on writing assignments in the unit, and presentation software may be used in project development. Additional technology integration suggestions are listed with lessons where appropriate.

Collaboration with Media Specialists

The exploration of science concepts can be considerably enhanced for students through the use of resources that bring science alive through pictures and stories.

Moreover, nonfiction resources support both teacher and student knowledge about systems. Teachers and media specialists can work together to collect such resources and have them available in the classroom or media center during the implementation of the unit. Some of the resources recommended in this unit may not be available at school libraries but may be found in public or university libraries, and specialists at these institutions can be very helpful as well in collecting listed resources and recommending additional ones. *Appendix D, Supplemental Readings for Students* lists both fiction and non-fiction texts for students.

Students should also be encouraged to utilize their library/media centers and to become acquainted with the librarians in their community for several reasons. First, libraries are complex systems of organizing information. The systems vary from one library to another, and technological access to the systems is constantly changing. Librarians serve as expert guides to the information maze and they are eager to assist library users. Secondly, the most important skill in using the library is knowing how to ask questions.

Students should learn that working with a media specialist is not a one-time inquiry or plea for assistance, but an interactive communication and discovery process. As the student asks a question and the librarian makes suggestions, the student will gain a better understanding of the topic and find new questions and ideas to explore. To maximize the use of library resources, the student should then discuss these new questions and ideas with the librarian. Learning to use the services of librarians and other information professionals is an important tool for lifelong learning.

2. Learning Centers

Learning centers can serve as useful tools throughout this unit, to give students more opportunities to explore the unit's topics and to deepen their understanding. A few suggestions for learning centers follow.

Models Center

At this center, various materials should be available to students for building models of objects such as houses and cars. For this center, the teacher needs to ensure that all materials provided will be used in a safe manner. This center may require teacher assistance and supervision.

Writing Center

Students may work at a writing center to strengthen their persuasive writing skills throughout the unit. They may use the writing center to revise and edit assignments given in the lessons, or to practice writing in response to other prompts provided. Copies of the Hamburger Model for Persuasive Writing, dictionaries, and thesauruses should be available at the writing center for students to reference.

Computer Center

A computer center may be used in several different ways during this unit. Students may use computer time for writing and editing or for working on a related research project for this unit. In addition, relevant websites involving Lyme disease, deer populations, and mathematical modeling may be bookmarked for students to visit and explore.

Map Analysis Center

At this center, maps should be available to students for further study. Blank maps should also be provided for students to make their own deer population maps. Students may conduct research regarding the deer populations and the related problems in various communities.

Outdoor Safety Center

Gather materials from the health department regarding safety precautions to take when working outside in wooded areas. Students may use the materials to develop presentations or other products to teach younger students about outdoor safety.

3. Teaching Models

A. The Taba Model for Concept Development

A variation of this concept development model is introduced in Lesson 5 of the unit. The concept of models is used as an organizer throughout the unit, with numerous questions and activities that explore students' understanding of the concept and the generalizations. This model is based on the work of Taba (1962).

The procedure as described below may be applied with various concepts and with students at various grade and ability levels.

Use the following questions to guide an introductory discussion about models. In groups, students should discuss the questions and record ideas on large sheets of paper for sharing with the class. Each section of the small group activity should be followed by a brief, whole-class discussion.

Brainstorm ideas about models and write down all responses.
- What words come to mind when you think about models?
- What are some examples of models? What is it about them that make them models?
- How do you decide whether something is a model?

Categorize the ideas that were written down, grouping them, and giving each group a title.
- How would you categorize these ideas into groups?
- What could you call each group? Why?

- Do all of your systems fall into groups? Might some of them belong in more than one group? How else might you group your ideas?
- What are some of the characteristics of models, based on the ideas you wrote?

Brainstorm a list of things that are not models.
- What are some examples of things that are not models?
- What evidence or proof do you have that these things are not models?
- How might you group the things that are not models? What can you call each of these models?
- How are the groups of things that are not models similar to or different from the groups of things that are models? Are there patterns to your groupings?

Make generalizations about *models*.

A generalization is something that is always or almost always true. What can you say about models that are always or almost always true? Use your examples and categories to guide your thinking and write several statements that are generalizations about models.

B. The Hamburger Model for Persuasive Writing

The purpose of the Hamburger Model is to provide students with a useful metaphor to aid them in developing a persuasive paragraph or essay. The model should be introduced by the teacher, showing students that the top bun and the bottom bun represent the introduction and conclusion of any persuasive writing piece. The teacher should note that the reasons given in support of the thesis statement are like the meat or vegetables in a hamburger, providing the major substance of the sandwich. Elaboration represents the condiments in a sandwich – the ketchup, mustard, and onions that enhance the sandwich's appeal – just as examples and illustrations enhance the appeal of a written persuasive writing piece.

Teachers should show students examples of hamburger paragraphs and essays and have students find the top bun, bottom bun, hamburger, and condiments. Students should have opportunities to evaluate the quality of the different components and the essay as a whole.

Teachers may ask students to construct their own hamburger paragraphs. After students have constructed their own paragraphs, teachers may use peer- and self-assessments to have students judge their own and one another's writing. This process should be repeated throughout the unit.

The Dagwood Model is an elaborated version of the Hamburger Model that uses the familiar metaphor of a sandwich to help students construct a paragraph or essay. Students begin by stating their point of view on the issue in question (the top bun). They then provide reasons, or evidence, to support their claim; they should try to incorporate at least three supportive reasons (the "patties"). Elaboration on the reasons provides additional detail (the "fixings"). A concluding sentence or paragraph wraps up the sandwich (the bottom bun). The Dagwood Model also asks students to introduce and refute other points of view.

With younger students, teachers may want to use a simpler model that includes only the top bun, bottom bun, and meat, without the elaboration. Students should be given the opportunity to master the simpler paragraph before moving on to developing more complex writing pieces. The following pages demonstrate both versions of the model in graphic format.

Hamburger Model for Persuasive Writing
(Primary Version)

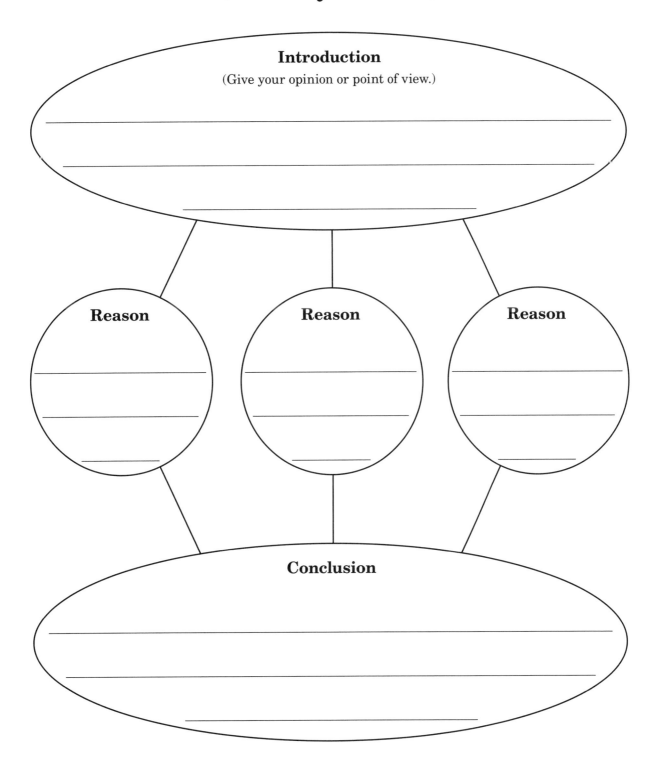

Introduction
(Give your opinion or point of view.)

_____

_____

_____

Reason

_____

_____

_____

Reason

_____

_____

_____

Reason

_____

_____

_____

Conclusion

_____

_____

_____

Hamburger Model for Persuasive Writing
(Version Including Elaboration)

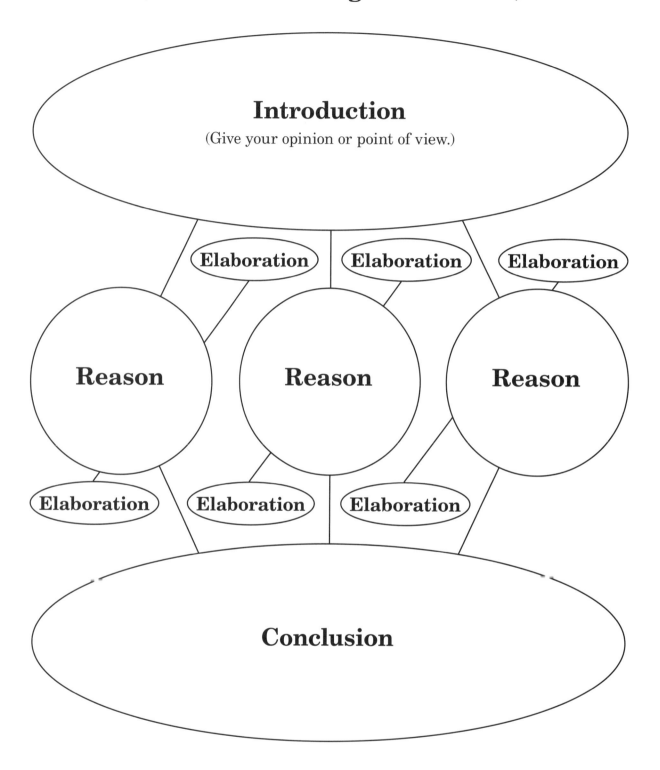

Introduction
(Give your opinion or point of view.)

Elaboration   Elaboration   Elaboration

Reason   **Reason**   **Reason**

Elaboration   Elaboration   Elaboration

Conclusion

Dagwood Model for Persuasive Writing

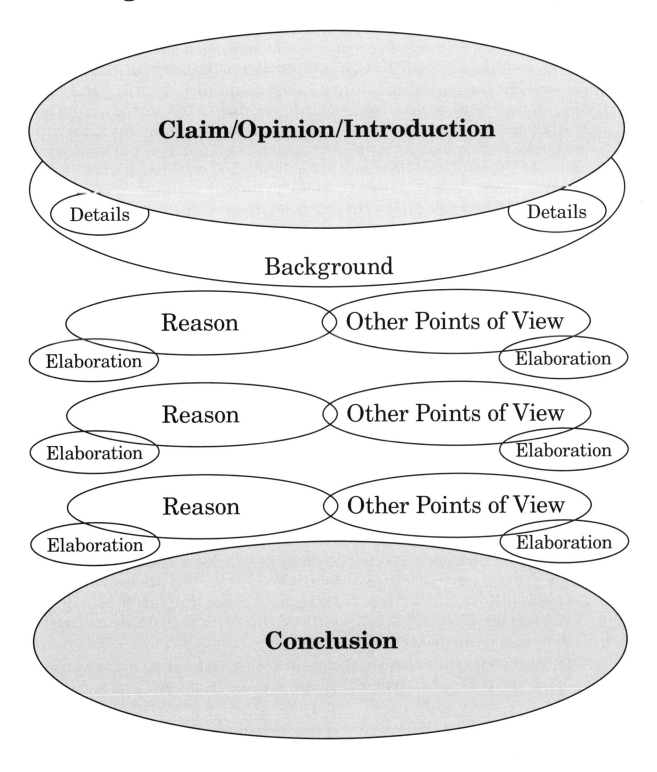

C. Elements of Reasoning

The reasoning strand used in this unit focuses on eight elements of thought identified by Richard Paul (1992). It is embedded in most lessons of the unit through questions, writing assignments, and research work. These elements of thought are the basic building blocks of productive thinking. Working together, they provide a general logic to reasoning. In document interpretation and listening, they help one make sense of the reasoning of the author or speaker. In writing and speaking, they enable authors or speakers to strengthen their arguments. In this unit for primary students, only some of the elements are introduced specifically, while others are only touched on through questions. This unit gives students an introduction to the elements that can be followed up in future units of study.

Students are often asked to distinguish between facts and opinions; however, between pure opinion and hard facts lie reasoned judgments in which beliefs are supported by reasons.

1. **Purpose, Goal, or End View:** We reason to achieve some objective, to satisfy a desire, or to fulfill some need. For example, if the car does not start in the morning, the purpose of my reasoning is to figure out a way to get to work. One source of problems in reasoning is traceable to defects at the level of purpose or goal. If our goal itself is unrealistic, contradictory to other goals we have, confused or muddled in some way, then the reasoning we use to achieve it is problematic. If we are clear on the purpose for our writing and speaking, it will help focus the message in a coherent direction. The purpose in our reasoning might be to persuade others. When we read and listen, we should be able to determine the author's or speaker's purpose.

2. **Question at Issue (or Problem to Be Solved):** When we attempt to reason something out, there is at least one question at issue or problem to be solved (if not, there is no reasoning required). If we are not clear about what the question or problem is, it is unlikely that we will find a reasonable answer, or one that will serve our purpose. As part of the reasoning process, we should be able to formulate the question to be answered or the issue to be addressed. For example, why won't the car start? Or, should libraries censor materials that contain objectionable language?

3. **Points of View or Frame of Reference:** As we take on an issue, we are influenced by our own point of view. For example, parents of young children and librarians might have different points of view on censorship issues. The price of a shirt may seem too low to one person while it seems high to another because of a different frame of reference. Any defect in our point of view or frame of reference is a possible source of problems in our reasoning. Our point of view may be too narrow, not be precise enough, unfairly biased, and so forth. By considering multiple points of view, we may sharpen or broaden our thinking. In writing and speaking, we may strengthen our arguments by

acknowledging other points of view. In listening and reading, we need to identify the perspective of the speaker or author and understand how it affects the message delivered.

4. **Experiences, Data, Evidence:** When we reason, we must be able to support our point of view with reasons or evidence. Evidence is important in order to distinguish opinions from reasons or to create a reasoned judgment. Evidence and data should support the author's or speaker's point of view and can strengthen an argument. An example is data from surveys or published studies. In reading and listening, we can evaluate the strength of an argument or the validity of a statement by examining the supporting data or evidence. Experiences can also contribute to the data of our reasoning. For example, previous experiences in trying to get a car to start may contribute to the reasoning process that is necessary to solve the problem.

5. **Concepts and Ideas:** Reasoning requires the understanding and use of concepts and ideas (including definitional terms, principles, rules, or theories). When we read and listen, we can ask ourselves, "What are the key ideas presented?" When we write and speak, we can examine and organize our thoughts around the substance of concepts and ideas. Some examples of concepts are freedom, friendship, and responsibility.

6. **Assumptions:** We need to take some things for granted when we reason. We need to be aware of the assumptions we have made and the assumptions of others. If we make faulty assumptions, this can lead to defects in reasoning. As a writer or speaker, we make assumptions about our audience and our message. For example, we might assume that others will share our point of view, or we might assume that the audience is familiar with the First Amendment when we refer to "First Amendment rights." As a reader or listener, we should be able to identify the assumptions of the writer or speaker.

7. **Inferences:** Reasoning proceeds by steps called inferences. An inference is a small step of the mind, in which a person concludes that something is so because of something else being so or seeming to be so. The tentative conclusions (inferences) we make depend on what we assume as we attempt to make sense of what is going on around us. For example, we see dark clouds and infer that it is going to rain; or we know the movie starts at 7:00—it is now 6:45, and it takes 30 minutes to get to the theater, so we cannot get there on time. Many of our inferences are justified and reasonable, but many are not. We need to distinguish between the raw data of our experiences and our interpretations of those experiences (inferences). Also, the inferences we make are heavily influenced by our point of view and assumptions.

8. **Implications and Consequences:** When we reason in a certain direction, we need to look at the consequences of that direction. When we argue and

support a certain point of view, solid reasoning requires that we consider what the implications are of following that path; what are the consequences of taking the course that we support? When we read or listen to an argument, we need to ask ourselves what follows from that way of thinking. We can also consider consequences of actions that characters in stories take. For example, if I do not do my homework, I will have to stay after school to do it; if I water the lawn, it will not wither in the summer heat.

Reasoning Wheel

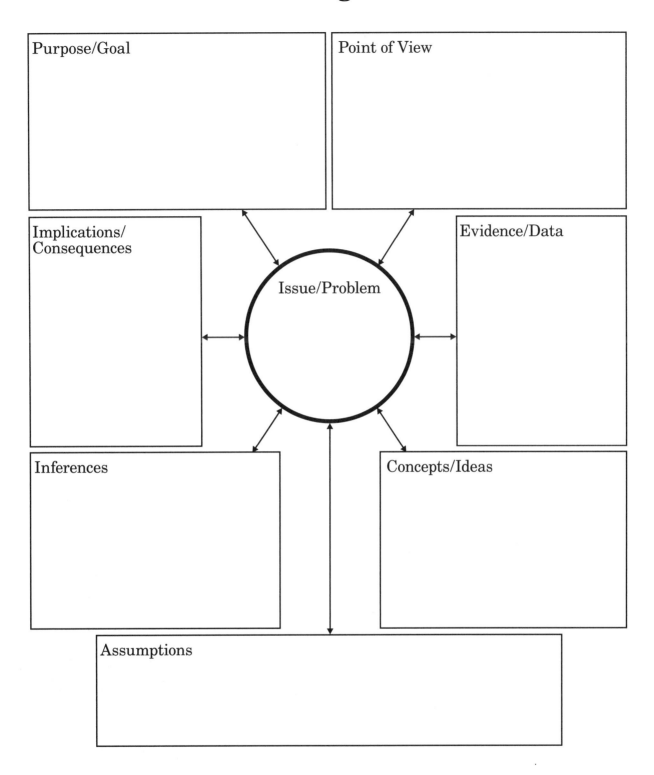

Adapted from Paul, R. (1992). *Critical thinking: What every person needs to survive in a rapidly changing world.* Sonoma, CA: Foundation for Critical Thinking.

Sample Reasoning Wheel

Purpose/Goal/End View

What is the purpose of discussing the deer population as it is related to the problem situation?

Point of View/Frame of Reference

From what perspective do you approach the issue? What interest groups or stakeholders may have different points of view on the topic of the most effective response? What might their viewpoints be?

Implications/ Consequences

What are the implications of this issue for the stakeholders?

Question/ Issue/Problem

How would an uncontrolled deer population affect the community?

Experience/ Evidence/Data

Research how the local agency that deals with wildlife would respond to this situation. What future research is planned in this area? What are the arguments that stakeholder groups give for and against the agency's response plan?

Inferences

After gathering data about the response plan for a related problem, describe what this involves for students in your school. Discuss and evaluate the arguments for and against the response plan. On what data are the stakeholders' arguments based? Have you changed your point of view after hearing and reading the facts? Explain.

Concepts/Ideas

How can a model help us in science? How are these ways evident when considering the issue of the deer population?

Assumptions

What assumptions emerge related to this issue? What assumptions might major stakeholders in this issue hold? Why do you think so?

Adapted from Paul, R. (1992). *Critical Thinking: What every person needs to survive in a rapidly changing world.* Sonoma, CA: Foundation for Critical Thinking.

4. Data Table Construction

Scientists communicate information through many methods, including organizational tools such as data tables. Designing data tables should be part of all students' experiences so that they can organize and communicate their findings clearly prior to drawing inferences from their collected data. *The National Science Education Standards* emphasize interpretation of data collected by students as early as the fourth grade. Younger students may be initially more comfortable with pre-collected sets of data.

Data refers to the measurements made when conducting experiments. Measurements of speed, distance, and temperature are examples of data. By organizing data into tables, the scientist can see patterns in the results.

In constructing data tables, there are some common conventions for facilitating communication between the writer and reader. For example, the independent variable is recorded in the left column and the dependent variable is recorded in the right column. Also, when repeated trials are conducted, the column for the dependent variable is divided into smaller columns so that data can be recorded for each repeated trial.

Source

Rezba, R. J., Sprague, C., Fiel, R. L. (2003). *Learning and assessing science process skills* (4th ed.). Dubuque, IA: Kendall/Hunt Publishing Company.

Title:					Average Measurement/Observation
Independent Variable	Dependent Variable				
	Trials				

Title The purpose of the experiment should be included in the title.

Independent Variable The independent variable should be clearly stated. This is the variable that is purposefully changed.

Dependent Variable The dependent variable should be clearly stated. If repeated trials are to be run, the columns under the dependent variable should be divided according to the number of trials run by the experimenter. This is the variable affected by the independent variable.

Average Measurement The average measurement and unit of measure should be listed, if applicable. Measurements should be organized either into ascending or descending order. A sample data table follows.

Sample Data Table

Title: Substances that repel deer

Independent Variable	Dependent Variable				Comments
	Trials				
soap					
vinegar					
rotten apple slices					
sour milk					
raw fish					

Appendices

4

The Concept of Models

A

The Concept of Models

(Taken from *The Guide to Key Science Concepts* by Beverly T. Sher)

A model is a simplified imitation of a thing or process. Models are used both in science and in the rest of human endeavor to better understand interesting things, to communicate their salient points to others, and to try out new ideas in a simplified way. To be useful, models must simulate the real-world thing that they represent in such a way as to match its behavior in a relevant way but make testing and prediction easier. One common simplification involves changes in size. For example, physical models of atoms are the size of children's blocks, but connect to each other at the same angles that the atoms use for bonding and are built to represent the relative sizes of the different kinds of atoms. The scale models of buildings used by architects are small enough to move around in a model landscape. Maps and globes are much smaller than the territory that they represent. Another common simplification is a change in timescale. Processes that occur too quickly to be amenable to study can be modeled at a slower speed; slow processes can be simulated at a higher speed. A third important type of simplification is the exclusion of all but relevant variables. Highway maps, for example, do not include the resource information found on geologic maps, as this information would only add visual confusion to the perplexities faced by a lost driver. A blueprint of a house specifies the dimensions for each room but doesn't specify the Formica pattern for the countertops in the kitchen. The blueprint is thus a relevant model for the house to the architect but fails to specify everything relevant for the interior decorator. By inclusion of only relevant variables, models can be tailor-made for every application.

Models can be physical, conceptual, or mathematical. An example of a physical model would be an architect's blueprint, a scale model of an airplane, a molecular model of a new compound, or a toy train set. Physical models are probably the easiest to understand, as there is something there to see, touch, and experience.

The second type of model is the conceptual model. Examples of conceptual models include thinking about the universe as a series of interconnected bubbles, or comparing the flow of traffic to the flow of a liquid through pipes. Conceptual models can be thought of as metaphors. The heart is like a mechanical pump; the situation in the Persian Gulf at the beginning of August 1990 was like the situation in Europe when Hitler reoccupied the Rhineland; the shape of blood vessels in the body is like the shape of the branches of a tree: all of these comparisons create conceptual models in which the salient features of the first member of each pair of items can potentially be used to understand the salient features of the second member of the pair.

The third type of model is the mathematical model. Mathematical models are used to mathematically simulate real-world processes: economic forecasting, weather prediction, and understanding the nature of the earliest moments of the universe all depend upon mathematical models. Such models consist entirely of

Source: Sher, B.T. (1993). *Guide to key and science concepts.* Williamsburg, VA: Center for Gifted Education.

mathematically defined relationships between the important variables involved in the behavior of the real-world system. Mathematical, physical, and conceptual models are used extensively by scientists and nonscientists alike.

In science, there is perhaps no better-known physical model than the double helical representation of the molecular structure of the DNA molecule. DNA (deoxyribonucleic acid) serves as the genetic blueprint for every organism that is built of cells (some viruses use RNA, a close relative of DNA, as their genetic material.) In the 1950s, James Watson and Francis Crick correctly deduced its structure with the help of X-ray crystallographic data provided by Maurice Wilkins and Rosalind Franklin. Using the X-ray data that showed that the molecule contained a helix as an essential element and their knowledge of the principles of chemical bonding, they built a physical model of the DNA molecule that satisfied all of the available data for its structure. The physical model (illustrated in Figure 2) immediately suggested to Watson and Crick a conceptual model that explained how DNA could direct the formation of copies of itself (illustrated in Figure 3). This conceptual model suggested that the DNA double helix first uncoils into single strands and that each strand is then copied by enzymes to generate two new identical double-helical DNA molecules, each of which contains one newly synthesized strand and one old strand. Subsequent experiments proved that this was indeed the case. This was of vast importance. A key feature of the properties of genetic material is that it has to be capable of self-replication: every time a cell divides, each new cell receives a complete copy of all of the genetic material that the parental cell possessed. A measure of the power of physical model-building in this example is the ease with which the physical model could be interpreted to enhance the understanding of the DNA molecule's function.

A second situation in science in which model-building has been critical is the case of the Big Bang, the event that describes the formation of the universe. Modeling the Big Bang has been a conceptual and mathematical exercise: this is not a system that lends itself to physical modeling, as physicists don't make new universes every day in the lab. Mathematical modeling, however, has led to predictions that are physically verifiable. For example, George Gamow first predicted that the Big Bang's initial explosion would leave a signature of sorts, a uniform background of microwave radiation present all through the universe. This mathematical prediction was made and refined in the 1950s; in 1964, Penzias and Wilson detected this microwave background using a radiotelescope, an experiment for which they later received the Nobel Prize. Thus, in this case, a conceptual model led to a mathematical model, which, in turn, led to an experimentally verifiable prediction. Much of modern physics has proceeded in exactly this way: conceptual and mathematical modeling tells the experimental physicists what to look for.

Yet another famous model in science is the concept of plate tectonics. It had been long known that there were similarities in geology between such widely separated regions as Africa and South America. In the early 20th century, Alfred Wegener formulated the hypothesis that the continents did not always occupy the same places that they do at present, but rather had drifted across the face of the globe over time. Thus, at one time, Africa and South America were physically linked, and during that time, geological phenomena had occurred that affected each of them identically: the history of their contact was therefore written in the rocks. For many years, Wegener's conceptual model was not accepted, as nobody could think of a mechanism to explain continental movements and the whole idea seemed rather far-fetched. Finally, though, a new form of data emerged, a way of testing rocks to see the magnetic field conditions present during their formation. The magnetic field of the Earth periodically reverses, and these reversals can be detected by study of the rocks and dated by using other methods of determining the age of the rocks. Paleomagnetism data provided unequivocal evidence for the movement of the continents, and Wegener was vindicated. Working through the implications of continental drift provided geologists with new explanations for old phenomena, including the locations of regions of volcanic and earthquake activity and the processes of mountain building. This conceptual model, the theory of plate tectonics, has thus greatly enriched the discipline of geology.

Model-building in science is not always as successful as it was in the three examples just mentioned, however. Many systems do not lend themselves to direct modeling. For example, the development of treatments and an eventual vaccine for AIDS has been hampered by the fact that only humans can be infected by HIV (the virus that causes AIDS) and can develop AIDS. In many disease states, the human condition can be directly simulated by an animal model. For example, diabetes was modeled in dogs: dogs whose pancreases have been removed become diabetic, and treatments for diabetes (such as injectable insulin) were tested in dogs before being tested in humans. In the case of AIDS, though, only an indirect animal model is available: monkeys have a virus similar to HIV known as SIV, and SIV infection in monkeys has been used to model HIV infection in humans. Because SIV and HIV are not identical, however, the testing of antiviral therapies and vaccine development has been heavily dependent on human volunteers, creating a situation rife with ethical problems and experimental limitations.

Even when a physical model is available, there can be other limitations. For example, the correspondence between the model system and the phenomenon being modeled may be inadequate or misleading. Biologists, for example, routinely test pharmacologically active compounds in animals before the FDA allows their use in people. This is supposed to protect humans from unpredicted side effects of new drugs and to ensure that the new drugs are effective for the condition for which they will be prescribed. Yet animals are not always perfect models for humans. The drug thalidomide, for example, was tested for safety in rats before it was approved as a tranquilizer in Europe for humans. There were no ill effects in rats, but it caused severe birth defects in humans who had been exposed to it as fetuses in utero. Imprecision of fit between the model and the real-life thing is a frequent limitation in the use of a model.

There are thus limits to the use of model-building in science. Understanding the uses and limitations of model-building is also important in other, nonscientific, areas. For example, the budget process that the Federal government goes through every year is highly dependent on the mathematical models of economists. Forecasting the federal deficit, for example, depends upon knowing how the economy will respond to events that could occur over the next budget period: war, decline in consumer confidence, increases in the price of energy, and so on. Curiously enough, forecasts emerging from the Executive Branch rarely match those from the Congress (particularly when the Congress and the White House are controlled by different political parties). This reflects several common limitations of model-building. The U.S. economy is highly complex and modeling it demands information on a very large number of variables. If the numbers fed into the computer models are incorrect, or reflect biased estimates, then the resulting projections will differ both from reality and from projections made by a group with different biases. The technical term that describes this problem is GIGO: garbage in, garbage out. In addition, the relationships among all of the variables governing the behavior of the U. S. economy are not all precisely known. Thus different economic models which treat these relationships differently will arrive at different predictions even when the same data are fed into them. It is very difficult to accurately model such complex systems: modeling the behavior of the Earth's atmosphere well enough to generate reliable weather forecasts is another case in point.

Weather forecasting is dependent on mathematical models that describe the behavior of the Earth's atmosphere. Weather data from hundreds of weather stations across the world are fed into huge computers every few hours, and the changes occurring over that time period are noted. Historical information about the patterns of change that have occurred in the past and theoretical descriptions of the behavior of different weather systems have been used to generate computer programs that relate the patterns of change occurring on a given day to each other and predict the likelihood of the occurrence of particular forms of weather in a particular place in the near future. The complexity of the system and the sheer number of variables and amounts of data involved combine to make long-term predictions less accurate than short-term predictions. Thus, forecasters can predict tomorrow's high temperature within a degree or three, but the extended forecast for the next month will only predict that the average temperature will be higher (or lower) than normal. The mathematical models that describe the behavior of the atmosphere are constantly being refined, and so weather prediction has become considerably more accurate over the past twenty or thirty years.

The use of physical, conceptual, and mathematical models is common in both scientific and nonscientific fields. Understanding how models are made, how they can be used, and the potential pitfalls of model systems is important for all educated adults. Many of these ideas can be addressed at a child's level, as described below.

Rationale for Teaching the Concept

The concept of models is potentially very easy to bring down to a child's level: they play with models every day. Their models are called toys. Blocks are models for building materials; a child that can build a stable tower has worked out some of the most basic engineering principles. Playing house allows children to model adult behavior and replicate the actions of their adult role models. Video games model such wonderful experiences as flying an airplane or killing evil creatures in the sewer system. The only type of model that children do not come to naturally is the mathematical model, and even this concept could be introduced to them in at least a simple form.

Working with models in a science context is a good way to bring scientific concepts down to a child's level. For primary age students, physical models are the most appropriate; for older students, conceptual and mathematical models can be worked in. If they have to use simple circuits to build a working radio-controlled model car, they'll pick up a better understanding of electricity along the way; building a better protection device for eggs than a styrofoam egg carton teaches some basic physics as well; building a better paper airplane teaches some aeronautics.

Making and using models can also help develop a child's scientific process skills. Being given materials and an unfamiliar problem to solve fosters higher-level thinking skills, if done well (fosters confusion and irritation, if done badly). Refining their models to fit new data helps them to understand that the scientific process is full of false starts and blind alleys but that progress can eventually be made nonetheless.

In building and using models in a scientific or technological context, children can develop an excellent feeling for what doing science is all about. The study of models lends itself to creativity, curiosity, and the ability to try new ideas and succeed in generating new insights.

Suggested Applications

Model building is accessible to every age group. The science may be worked in around the model-building, or vice versa. The strategy of problem-based learning can easily be employed to teach this concept. If the students were given a challenge (figure out how to design a paper airplane that could stay in the air for 20 seconds after being thrown from a third-floor window, for example) they could be encouraged to go find the relevant science on their own. Facilitated by a teacher, this would become an exercise in the practical application of the principles of aerodynamics to model planes. The possibilities listed for the younger grades could also be used for older children in greater detail and depth.

For grades K–2:
- Blueprints: making and reading
- Maps and mapmaking
- Paper airplane contests
- Terraria
- Aquariums
- Building bridges

For grades 3–5:

- Dollhouse construction (with working plumbing and electricity)
- Building a better model boat
- Robotics with electronic Legos
- Designing a Mars colony

For grades 6–8:

- The egg drop experiment: building better packaging
- Geologic maps: making and using maps
- Orienteering
- Fun with molecular models: both physical and computer-generated
- Simple mathematical modeling of physics concepts; tie it in with physical experiments performed first

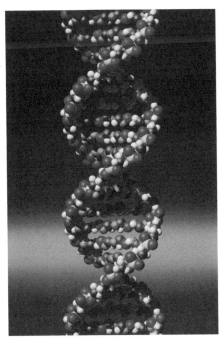

Figure 1 The figure shows a space-filling model of the DNA double helix. © Digital Art/Corbis.

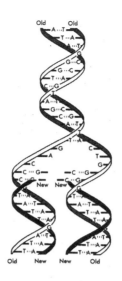

The replication of DNA.

Figure 2 This figure shows a diagram of the processes of DNA replication. Fig. 9.7, p. 212 from *Molecular Biology of the Gene* by James D. Watson. Reprinted by permission of Addison Wesley Longman Publishers, Inc.

R e f e r e n c e s

Dethier, V. (1962). *To know a fly.* San Francisco, CA: Holden-Day.

Judson, H. F. (1980). *The search for solutions.* NY: Holt, Rinehart, and Winston.
   (A very readable account of the scientific process; model systems for a number of scientific areas are discussed.)

Simon, H. A. (1991). *Models of my life.* NY: Basic Books.
   (The autobiography of Herbert Simon, who won the Nobel Prize in economics and pioneered in the field of artificial intelligence: a good reference for mathematical models; this is another in the Sloan Foundation series of scientific autobiographies.)

Sullivan, W. (1974). *Continents in motion: The new earth debate.* NY: McGraw-Hill Book Co.
   (A readable account of the history of the theory of plate tectonics.)

Watson, J. D. (1972). *The double helix: A personal account of the discovery of the structure of DNA.* NY: Atheneum.

Suggested Mini-Lessons

B

gested Mini-Lessons

The following suggested mini-lessons may be points at which review of concepts is necessary for students to grasp the intent of the lessons in the unit more easily.

Lesson	Suggested Mini-Lesson	Mini-Lesson Online Links
Lesson 3: Learning About Lyme Disease	Use this lesson plan to provide additional content information about parasite-borne diseases.	http://www.nationalgeographic.com/xpeditions/lessons/18/g912/parasites.html
Lesson 6: Maps as Models	Use this lesson plan to assist students in constructing topographical maps.	http://www.nationalgeographic.com/xpeditions/lessons/01/g68/dogstails.html
Lesson 7: The Niche of Deer	Use this lesson plan for a simulation in deer population management.	http://sftrc.cas.psu.edu/LessonPlans/Wildlife/DeerIssues.html
Lesson 10: Sampling/ Estimating	Use this lesson plan to help students estimate the size of a population by using statistical procedures.	http://www.pbs.org/teachersource/mathline/lessonplans/pdf/msmp/somethingfishy.pdf
Lesson 12: Exponential Growth Model	Use the lesson plan to introduce students to exponential functions.	http://www.learner.org/channel/workshops/algebra/workshop6/lessonplan.html
Lesson 12: Exponential Growth Model	Use this lesson plan as introduction to mathematical modeling.	http://www.learner.org/channel/workshops/algebra/workshop8/lessonplan2.html
Lesson 17: Humans vs. Deer	Use this website for resources related to debating.	http://www.educationworld.com/a_lesson/lesson/lesson304.shtml

Interdisciplinary Connections

Interdisciplinary Connections

These interdisciplinary connections activities may be used in a variety of ways. They may be:

- Assigned to students as extension activities;
- Used for enrichment purposes in classroom learning centers; or
- Included as supplementary activities for the unit.

Interdisciplinary Connections

Connecting to . . . *Mathematics*

- Locate examples of population growth models for various cities, counties, and states in the United States. Research how such models are used in community planning.
- Conduct a survey of your classmates about a topic of your choice. After collecting the data, use spreadsheet software to design related charts and graphs showing your results.

Connecting to . . . *Language Arts*

- Use the Hamburger Model or Dagwood Model as a guide to help you organize your response to the following essay prompt: How should people protect themselves from parasitic insects when working or playing outside?
- Write an essay comparing the deer population in your locality now to 50 years ago, 100 years ago, etc. In the essay, address the potential reasons for the change in the deer population.

Connecting to . . . *Social Studies*

- Research the impact of commercial and residential development on the populations of various animals in your community. Include information regarding the measures taken to protect the animals and the environment.
- Insect-borne diseases have resulted in many deaths at various times in history. Research an example of such a disease and find related information about the impact in a given part of the world, including the death toll, measures taken to protect humans, and the occurrence of the disease at this point in time.

Supplemental Readings for Students

D

Supplemental Readings for Students

These fiction and non-fiction books may be used to supplement *Animal Populations: A Study of Physical, Conceptual, and Mathematical Models*. Fiction books could be used in the language arts class as students complete the science unit. The non-fiction books are good resources for student research activities related to the unit.

Fiction	Non-Fiction
My Side of the Mountain by Jean Craighead George Penguin Young Readers Group, 2004 ISBN: 0142401110 (subject: hunting)	*Reptile Rescue* by Peggy Thomas Lerner Publishing Group, 2000 ISBN: 0761332324 (subject: protect and monitor reptiles)
Tracker by Gary Paulsen Simon & Schuster Children's Publishing, 1995 ISBN: 0689804121 (subject: deer hunting)	*Animals Among Us: Living with Suburban Wildlife* by Fran Hodgkins Shoe String Press, Incorporated, 2000 ISBN: 0208024786
The Yearling by Marjorie Kinnan Rawlings Simon & Schuster Children's Publishing, 1988 ISBN: 0020449313 (subject: deer)	*Lyme Disease* by Alvin Silverstein, Virginia B. Silverstein, Laura Silverstein Nunn Scholastic Library Publishing, 2000 ISBN: 0531165310
Rescue of Josh McGuire by Ben Mikaelse Hyperion Books for Children, 1993 ISBN: 1562825232	*Lyme Disease (Diseases and Disorders Series)* by Gail B. Stewart Thomson Gale, 2003 ISBN: 1560069074
Brian's Hunt by Gary Paulsen Random House Children's Books, 2005 ISBN: 0553494155	*Deer, Moose, Elk and Caribou* by Deborah Hodge Kids Can Press, Limited, 2004 ISBN: 1550746677
My Side of the Mountain by Jean Craighead George Penguin Young Readers Group, 2004 ISBN: 0142401110 (subject: hunting)	*Reptile Rescue* by Peggy Thomas Lerner Publishing Group, 2000 ISBN: 0761332324 (subject: protect and monitor reptiles)

Suggested
Rubrics

E

Suggested Rubrics

The rubrics provided in this section may be used for assessment purposes when teaching the unit. The rubrics include:

- Experimental Design Rubric
- Connections to Systems Concept Rubric
- Oral Presentation Rubric
- Persuasive Writing Rubric

A source of rubrics for use specifically in the science classroom is:

Lantz, H. B. (2004). *Rubrics for assessing student achievement in science grades K–12*. Thousand Oaks, CA: Corwin Press, Inc.

Experimental Design Rubric

Criteria	Strong Evidence 2	Some Evidence 1	No Evidence 0	Pre	Post
States **PROBLEM** or **QUESTION**.	Clearly states the problem or question to be addressed.	Somewhat states the problem or question to be addressed.	Does not state the problem or question to be addressed.		
Generates a **PREDICTION** and/or **HYPOTHESIS**.	Clearly generates a prediction or hypothesis appropriate to the experiment.	Somewhat generates a prediction or hypothesis appropriate to the experiment.	Does not generate a prediction or hypothesis.		
Lists experiment steps.	Clearly & concisely lists four or more steps as appropriate for the experiment design.	Clearly & concisely lists one to three steps as appropriate for the experiment design.	Does not generate experiment steps.		
Arranges steps in **SEQUENTIAL** order.	Lists experiment steps in sequential order.	Generally lists experiment steps in sequential order.	Does not list experiment steps in a logical order.		
Lists **MATERIALS** needed.	Provides an inclusive and appropriate list of materials.	Provides a partial list of materials needed.	Does not provide a list of materials needed.		
Plans to **REPEAT TESTING** and tells reason.	Clearly states a plan to conduct multiple trials, providing reasoning.	Clearly states a plan to conduct multiple trials.	Does not state plan or reason to repeat testing.		
DEFINES the terms of the experiment.	Correctly defines all relevant terms of the experiment.	Correctly defines some of the relevant terms of the experiment.	Does not define terms, or defines terms incorrectly.		
Plans to **MEASURE**.	Clearly identifies plan to measure data.	Provides some evidence of planning to measure data.	Does not identify plan to measure data.		
Plans **DATA COLLECTION**.	Clearly states plan for data collection, including note-taking, the creation of graphs or tables, etc.	States a partial plan for data collection.	Does not identify a plan for data collection.		
States plan for **INTERPRETING DATA**.	Clearly states plan for interpreting data by comparing data, looking for patterns and reviewing previously known information.	States a partial plan for interpreting data.	Does not state plan for interpreting data.		
States plan for drawing a **CONCLUSION BASED ON DATA**.	Clearly states plan for drawing conclusions based on data.	States a partial plan for drawing conclusions based on data.	Does not state plan for drawing conclusions.		
					TOTAL SCORE:

Adapted from Fowler, M. (1990). The diet cola test. *Science Scope, 13(4)*, 32–34.

Connections to Systems Concept Rubric

Target Skills	Novice	Developing	Proficient	Exemplary
Connection to Systems Concept	• Student has little or no knowledge of systems vocabulary.	• Student exhibits knowledge of systems vocabulary.	• Student attempts to use systems vocabulary; the use is sometimes inaccurate.	• Student uses systems vocabulary accurately.
	• Student treats the concept of systems and scientific processes separately.	• Student recognizes links between the concept of systems and scientific processes.	• Student links the concept of systems with scientific processes.	• Student links the concept of systems and scientific processes in novel ways.
	• Student makes no connection between the problem resolution and the concept of systems.	• Student demonstrates limited ability to connect problem resolution with the concept of systems.	• Student demonstrates the ability to connect problem resolution with the concept of systems.	• Student recognizes problem resolution and systems connections beyond the scope of the unit problem.

Oral Presentation Rubric

Target Skills	Novice	Developing	Proficient	Exemplary
Oral Communication Skills	• The speaker's purpose is unclear.	• The speaker's purpose needs additional clarification.	• The speaker's purpose is clear.	• The speaker's purpose is clear and fully developed.
	• Student maintains little or no eye contact with audience.	• Student sometimes maintains eye contact with audience.	• Student mostly maintains appropriate eye contact with audience.	• Student consistently maintains appropriate eye contact with audience.
	• Student articulation is unclear and difficult to understand.	• Student articulation is generally clear but may not always be correct.	• Student articulates clearly and correctly.	• Student articulates clearly, correctly, and precisely.
	• Student does not use appropriate volume.	• Student is sometimes difficult to hear.	• Student uses appropriate volume most of the time.	• Student uses appropriate volume and considers audience size and room capacity.

continued

Oral Presentation Rubric (Continued)

Target Skills	Novice	Developing	Proficient	Exemplary
Oral Communication Skills (continued)	• Distracters in language and body movement overwhelm the presentation.	• Minimal distracters in language and body movement are present.	• Distracters in language and body movement are absent.	• Students uses language and body movement to enhance, not distract from, the presentation.
	• Student word choice is inappropriate and imprecise.	• Student word choice is inappropriate and/or imprecise.	• Student word choice is appropriate and precise.	• Student word choice is precise and sophisticated.
Organization Skills	• Little organization is evident in presentation.	• Limited organization is evident in presentation.	• Information is presented in an organized sequence.	• Information is presented in an organized and engaging sequence.
	• The presentation shows little evidence of planning.	• The presentation shows some evidence of planning.	• The presentation shows thoughtful planning.	• The presentation shows thoughtful planning and careful construction.
Visual Display Skills	• Visual aid(s) detract from the presentation.	• Visual aid(s) are not appropriately incorporated into the presentation.	• Visual aid(s) complement presentation information.	• Visual aid(s) enhance and reinforce presentation information.
Problem Resolution Skills	• The problem resolution does not consider stakeholder perspectives.	• The problem resolution makes some reference to stakeholder perspectives.	• The problem resolution considers a limited number of stakeholder perspectives.	• The problem resolution considers multiple stakeholder perspectives.
	• There is no connection made between the problem resolution and the information discovered through research.	• Some connection is made between the problem resolution and the information discovered through research.	• Obvious connections are made between the problem resolution and information discovered through research.	• Sophisticated connections are made between the problem resolution and information discovered through research.

Persuasive Writing Scoring Rubric

Claim or Opinion

0   No clear position exists on the writer's assertion, preference, or view, and context does not help to clarify it.

2   Yes or no alone or writer's position is poorly formulated, but reader is reasonably sure what the paper is about based on context.

4   *Meets expectations:* A clear topic sentence exists, and the reader is reasonably sure what the paper is about based on the strength of the topic sentence alone.

6   *Exceeds expectations:* A very clear, concise position is given and position is elaborated with reference to reasons; multiple sentences are used to form the claim. Must include details that explain the context.

Data or Supporting Points

0   No reasons are offered that are relevant to the claim.

2   One or two weak reasons are offered; the reasons are relevant to the claim.

4   At least two strong reasons are offered that are relevant to the claim.

6   *Meets expectations:* At least three reasons are offered that are relevant to the claim.

8   *Exceeds expectations:* At least three reasons are offered that are also accurate, convincing, and distinct.

Elaboration

0   No elaboration is provided.

2   An attempt is made to elaborate on at least one reason.

4   More than one reason is supported with relevant details.

6   *Meets expectations:* Each reason (3) is supported with relevant information that is clearly connected to the claim.

8   *Exceeds expectations:* The writer explains all reasons in a very effective, convincing, multi-paragraph structure.

Conclusion

0   No conclusion/closing sentence is provided.

2   A conclusion/closing sentence is provided.

4   *Meets expectations:* A conclusion is provided that revisits the main ideas.

6   *Exceeds expectations:* A strong concluding paragraph is provided that revisits and summarizes main ideas.

5

References
and Resources

References

Alexander, M. M. (1958). The place of aging in wildlife management. *American Scientist 46,* 123–137.

American Association for the Advancement of Science (1990). *Science for all Americans.* New York: Oxford University Press.

Cassidy, J. (1993.) *Earthsearch: A kid's geography museum in a book.* Palo Alto, CA: Klutz Press.

Cornicelli, L., Woolf, A., & Roseberry, J. L. (1993). Residential attitudes and perceptions toward a suburban deer population in Southern Illinois. *Transactions of the Illinois State Academy of Science, 86,* 23–32.

Cothron, J. H., Giese, R. N., & Rezba, R. J. (2005). *Science experiments and projects for students.* Dubuque, IA: Kendall/Hunt Publishing Company.

Cothron, J. H., Giese, R. N., & Rezba, R. J. (2000). *Students and research: Practical strategies for science classrooms and competition.* Dubuque, IA: Kendall/Hunt Publishing Company.

Craven, S. & Hygnstrom, S. (1991). *Controlling deer damage in Wisconsin.* Madison, WI: Cooperative Extension Publications, Publication G3083.

Ellingwood, M. R. & Caturano, S. L. (1989). *An evaluation of deer management options.* Hartford, CT: Connecticut Department of Environmental Protection.

Garrett, L. (1994). *The coming plague: Newly emerging diseases in a world out of balance.* New York: Farrar, Straus, & Giroux.

Gerlach, D. Atwater, S., & Schnell, J. (1994). *The wildlife series: Deer.* Mechanicsburg, PA: Stackpole Books.

Higgins, A. (1996, February 8). When deer invade the gardens: Suburban gardens are under siege. Strategic planning may win the war. *The Washington Post,* T8.

Jefferson, R. (Ed.). (1995, November). *Virginia Wildlife, 56* (11).

Kantor, F. S. (1994, September). Disarming Lyme disease. *Scientific American.* 34–39.

Karlen, A. (1995). *Man and microbes: Disease and plagues in history and modern times.* New York: Putnam.

Krebs, C. J. (1964). The lemming cycle at Baker Lake, Northwest Territories, during 1959–62. *Tech. Paper No. 15, Arctic Institute of North America.*

Maddux, H. C. (Ed.) (1986). *The Challenge of the Unknown: Teaching Guide.* New York: W. W. Norton & Company.

Ostfeld, R. S., (1997, July–August). The ecology of Lyme-disease risk. *American Scientist, 85,* 338–346.

Ostfeld, R. S., Jones, C. G., & Wolff, J. O. (May 1996). Of mice and mast: Ecological connections in eastern deciduous forests. *BioScience, 46* (5), 323–330.

Peyser, M. (1996, September 9), Backyard bears. *Newsweek,* 76–77.

Schmitt, B. (2005). *Your Child's Health: The parent's guide to symptoms, emergencies, common illnesses, behavior, and school problems (rev. ed.).* New York: Bantam Books.

Schneider, D. & Kuser, J. (1989, January/February). Suburbia: Too many deer or too many people? *New Jersey Outdoors,* 28–32.

Stein, S. (1993). *Noah's garden.* Boston: Houghton Mifflin Company.

Resources

Websites

The Department of Health and Human Services Centers for Disease Control and Prevention

This site provides information about prevention, treatment, symptoms, and statistics about Lyme disease. Statistical information includes a map of reported cases across the United States, graphs, and charts. Also included is a link to a tick prevention handbook.
http://www.cdc.gov/ncidod/dvbid/lyme/index.htm

The Lyme Disease Foundation

This site provides pictures, information about vaccines, legislation connections, and a children's corner on their website. Examine the picture gallery to see pictures of ticks, rashes, other symptoms, etc.
http://www.lyme.org/

Maryland's State Department of Natural Resources

This site provides an overview of the problems that exist between humans and deer. Included is a link to an appendix that describes Maryland's history of deer management.
http://www.dnr.state.md.us/wildlife/deerandhumans.html

Math Teacher Links

This site provides an online introduction tutorial to TI-82 graphing calculators. Through the interactive tutorial, students can learn the keyboard functions, graphing functions, etc.
http://mtl.math.uiuc.edu/non-credit/basic82/

National Geographic Map Machine

National Geographic sponsors a map machine that allows you to access their renowned map collection, including topographic maps. Maps are also available for purchase.
http://plasma.nationalgeographic.com/mapmachine/

The Pennsylvania Game Commission

The Pennsylvania Game Commission dedicates a portion of its website to a deer management program. The program stresses that the community needs to work together to combat the conflict between deer and humans and that there is no simple solution.
http://www.pgc.state.pa.us/pgc/cwp/view.asp?a=465&q=161556

The Quality Deer Management Association

This is an organization that strives to balance deer herds within the community context. They provide deer population management strategies and provide articles about data collection, deer biology, hunting management, etc.
http://www.qdma.com/

Modular Unit

- *Populations and Ecosystems,* Grades 6–8
  FOSS Unit: Full Option Science System
  Developed by Lawrence Hall of Science Center for Curriculum Innovation
  University of California, Berkeley
  Lawrence Hall of Science #5200
  Berkeley, CA 94720-5200
  http://lhsfoss.org/

Index